In the Land of the Wild Onion

In the Land of the Wild Onion

Travels Along Vermont's Winooski River

Charles Fish

University of Vermont Press
Burlington, Vermont

PUBLISHED BY UNIVERSITY PRESS OF NEW ENGLAND
HANOVER AND LONDON

University of Vermont Press
Published by University Press of New England,
One Court Street, Lebanon, NH 03766
www.upne.com

Printed in the United States of America
5 4 3 2 1

Library of Congress Cataloging-in-Publication Data

Fish, Charles.
In the land of the wild onion : travels along Vermont's Winooski River / Charles Fish.
p. cm.
Includes bibliographical references.
ISBN-13: 978-1-58465-550-3 (cloth : alk. paper)
ISBN-10: 1-58465-550-X (cloth : alk. paper)
1. Winooski River (Vt.)—Description and travel. 2. Winooski River Valley (Vt.)—Description and travel. 3. Fish, Charles—Travel—Vermont—Winooski River Valley. 4. Winooski River Valley (Vt.)—Social life and customs. 5. Winooski River Valley (Vt.)—Environmental conditions. 6. Outdoor life—Vermont—Winooski River Valley. I. Title.
F57.W73F57 2006
917.43'17—dc22 2006001966

green press INITIATIVE University Press of New England is a member of the Green Press Initiative. The paper used in this book meets their minimum requirements for recycled paper.

For Eleanor

with love and gratitude

Winoskik was the name of an Abenaki village near the mouth of the river: "*Winoskik* means 'At Wild Onion Land,' probably because the land nearby is suitable for the growth of the onionlike wild leek *Allium canadense.*"

—WILLIAM A. HAVILAND AND MARJORY W. POWER,
The Original Vermonters: Native Inhabitants, Past and Present

Water, odorless, colorless, tasteless, transparent liquid that is colorless in small amounts but exhibits a bluish tinge in larger quantities. It is the most familiar and abundant liquid on earth. In solid form (ice) and liquid form it covers about 70% of the earth's surface. It is present in varying amounts in the atmosphere. Most of the living tissue of man is made up of water; it constitutes about 92% of blood plasma, about 80% of muscle tissue, about 60% of red blood cells, and over half of all other tissues. It is an important component as well of the tissues of most other living things. —*The Columbia Encyclopedia,* 5th ed.

Yes, as every one knows, meditation and water are wedded for ever.

—HERMAN MELVILLE, *Moby-Dick*

Contents

Preface

Rivers seldom run straight, and neither does my book. Canoeing and hiking, I twice traced every bend of the Winooski River from the headwaters to Lake Champlain. But the story takes other turns as well. In fact, more pages by far are given to the natural and human history of the valley than to my days on the water. Since what we see is conditioned largely by what we know, I wanted to learn as much as possible of the many ways other people had come to know the valley and had made it their home.

I kept a travel log, I interviewed people, I read as widely as I could. Topical and travel chapters are interspersed with some regularity as I make my way down the river. Thanks to my many sources, my journey was a voyage of discovery much richer than it would have been without their help. If at times the discovery introduces unfamiliar terms and daunting details, I can only ask the reader to paddle on with me. Until we reach the lake, there is always another bend in the river, another realm to be explored.

Some topics of current interest will have taken on new twists and bends before this story reaches print, but time should not alter their character and significance.

Acknowledgments

My heartfelt thanks to my wife, Eleanor, best of readers, who often saw the better way and saved me from many infelicities. Thanks also to my editor, Richard Pult, for his many helpful suggestions.

Some thirty-five people appear as major or minor characters in the book. They opened my eyes. If I do not mention them here, it is because my debt will be obvious when they come on stage. There are others who do not appear but who provided advice and information for which I am also deeply grateful. I list them here with apologies to any I may have overlooked.

From the Vermont Agency of Natural Resources: Marjorie Gale and Diane Vanacek (Office of the State Geologist); Brian Chipman, Lawrence

Garland, Chet MacKenzie (Department of Fish and Wildlife); Karen Bates, Timothy Clear, Jeffrey Cuetto, Michael Kline, Peter LaFlamme, Jerome McArdle, Thomas Willard (Water Quality Division); Paul Olander (Wastewater Division).

Archaelogists: Douglas Frink (Archaeology Consulting Team, Inc.); William Haviland, James Petersen, Marjory Power (University of Vermont); Giovanna Peebles (Vermont Division for Historic Preservation).

Librarians: Christopher Burns, Prudence Doherty (University of Vermont Special Collections); Paul Carnahan and staff (Vermont Historical Society Library); Jerry Carbone and staff (Brooks Memorial Library, Brattleboro, Vermont).

Other: The Rev. Peter Anderson (Jericho Congregational Church); Dale Arengo (Champlain Mill); David Bryan (Ethan Allen Homestead); Brian Carlson, Jennifer Ely (Winooski Valley Park District); Terry Cecchini (Green Mountain Power Corp.); Bruce Chapell, Ellen Sivret (Natural Resources Conservation Service); Kinny Connell (Friends of the Mad River); Eric Derleth, Nicholas Staats (United States Fish and Wildlife Service); Carol Hulse (mill information); Laura Krawitt (Heritage Winooski); Allen Matthews, Gwyneth Harris (The Center for Sustainable Agriculture, University of Vermont); Ray Pecor (Lake Champlain Transportation Co.); Stephan Syz (Montpelier Conservation Commission); Kitty Werner (Mad River information).

Any errors are of course my own.

In the Land of the Wild Onion

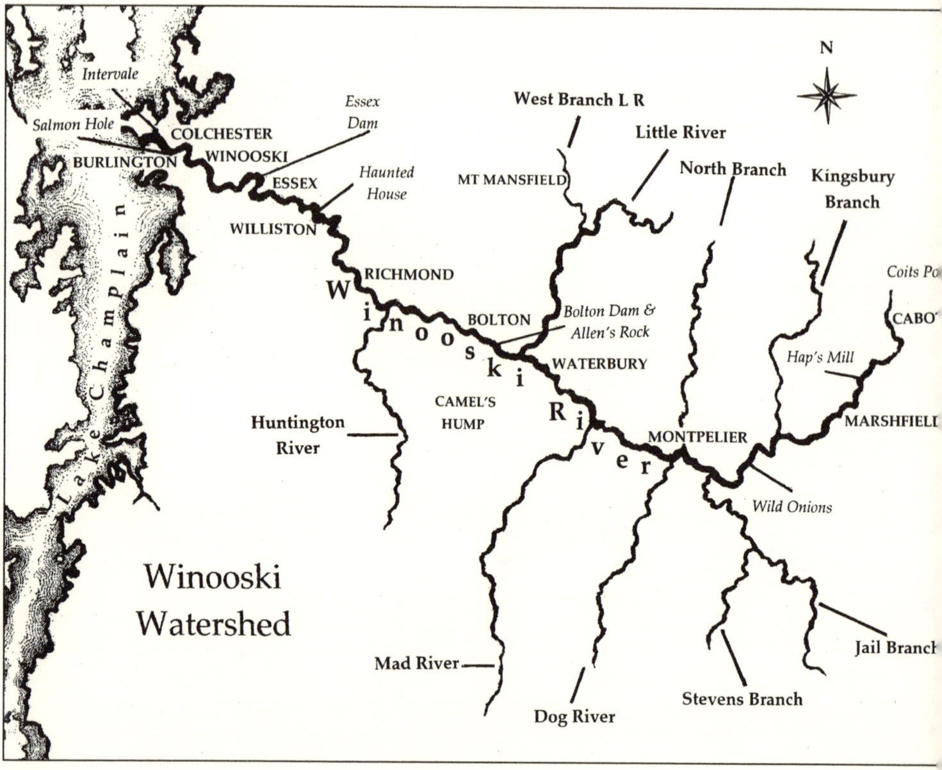

G. HEGBERG BLOMG

I

My River, My Valley

Since the first human eye saw a leaf in Devonian sandstone and a puzzled finger reached to touch it, sadness has lain over the heart of man. By this tenuous thread of living protoplasm, stretching backward into time, we are linked forever to lost beaches whose sands have long since hardened into stone. The stars that caught our blind amphibian stare have shifted far or vanished in their courses, but still that naked, glistening thread winds onward. No one knows the secret of its beginning or its end. Its forms are phantoms. The thread alone is real; the thread is life.

—LOREN EISELEY, *The Firmament of Time*

The uniqueness of an individual creature is inherent, not in its physical or behavioral anomalies, but in its life. Its life is not its "life history," the typical cycle of members of its species from conception to reproduction to death. Its life is all that happens to it in its place. Its wholeness is inherent in its life, not in its physiology or its biology. This wholeness of creatures and places together is never going to be apparent to an intelligence coldly determined to be empirical or objective. It shows itself to affection and familiarity.

—WENDELL BERRY, *Life Is a Miracle: An Essay Against Modern Superstition*

As a youngster I lived near the river just a few miles from where the Green Mountains abruptly give way to the Champlain Lowlands, a marked geological transition. As a child, however, I thought as a child; I didn't think geologically. I lived on intimate terms with the features of earth and water, and only later, under the tutelage of those whose lives were devoted to the contemplation of such things, did I come to glimpse a different unity from the one experienced in childhood. To a Midwesterner the landscape would not seem flat, but to a Vermonter the knobs and hills were nothing compared to the steeper grades that rose but a few miles to the east and that in turn were succeeded quickly by the sharp slopes and cliffs of the mountains.

Essex Junction, my village, sat on sandy terraces with the Winooski River as its southern boundary. When I played on the sand bank behind our School Street house, I felt the grains between my toes, hot from the sun. Where the sand came from was not a question that occurred to me.

No one told me that if I had been standing there some 18,000 years ago, the probable peak of the last glacier, I would have been covered by one to two miles of glacial ice and debris; if 13,000 years ago, I would have been hundreds of feet under the water of so-called Lake Vermont; and if 11,000 years ago, I would still be very wet but with the briny water of the Champlain Sea. I was ignorant of the geological processes by which streams and lakes and seas produce sand for little boys to play in. And clay, too, for when I was a little older I was given an ancient Stevens .22 to hunt squirrels in the oaks at the other side of the village, and on my way I passed the clay pits of the brickyard where since 1867 the Drury family made bricks still found far and wide.

If not hunting, I might be fishing in the river, either in the still waters above the dam or in the eddies and pools behind the power plant. The river often ran dark with silt, depositing it behind the dam and elsewhere in the quiet waters downstream as it passed the Whitcomb farm where plowing turned up arrowheads or "projectile points," as the archaeologists have taught us to say. Farther on, it meandered even more slowly through the ancient delta of the Intervale, an agricultural and wildlife oasis increasingly surrounded by urban development. And the river ran with sewage, too, known to all but of little concern to the young fisherman except that his mother was not eager to cook the fish he caught.

Somehow these things and more were all of a piece in the life of a child. There was nothing that cried out for a more penetrating or more comprehensive view of the phenomena of my little world—the sand, clay, oaks, fish, dam, generating plant, arrowheads, silt, and sewage—and yet they came together; they came together in the happy egocentricity of childhood. I was the center in which all things met. Although I may have been aware that they belonged to orders independent of me, it was their vivid particularity that mattered, not their general categories under the aspect of science. At the bend in the river, the boulder against which I leaned was uniquely that boulder, always there, not just one of millions of erratics deposited on the retreat of the last glacier; the smallmouth bass that took my nightcrawler was my bass, the bass of the moment, not *Micropterus dolomieui.* There is something, however, that spans the decades, that binds the memory of a boyhood well spent with the rediscovery of river and valley in recent years, and that something is love, love of place. Even now, after many trips north from my home in southern Vermont, my heart beats a little faster when I crest the high land above Barre on Interstate 89, spot the tip of Camel's Hump far to the west, and let my eye traverse the ridges that the river has cut over time.

Love and knowledge—a perfect union if life were perfect, as it is not. Love of place is acquired without trying and, unlike the more vexing human love, requires no reciprocity. It is a gift of intimacy. Bad luck—an unresponsive nature or unhappy experience—can block or sour it, but for the fortunate, and they are many, this quiet love grows with the years. It is otherwise with knowledge. Intimacy may provide material and may move the mind to grasp what the heart holds, but to know is to see with trained eyes, to think at one remove from the immediacy of experience.

And so it was that when adult inquiry joined childhood memory to explore the Winooski Valley, I soon discovered how much there was to see and how ill-prepared I was to see it. Imagine wandering through a castle of vast dimensions—every room leading to untold others—and that castle not only an artifact of the present but with a history as well. Or draw a circle on the ground a yard in diameter and imagine a diminishing cone projected to the earth's core. To understand deeply all the physical phenomena in the cone—the plant and animal life, the soil formation, the deep earth geology—would take a lifetime or more.

The word "valley" brings to mind hills and mountains sloping to water, and indeed for most of the way that is just what I saw when I followed the Winooski River from its source to Lake Champlain. What I call the valley is, more precisely, the watershed or, even more prosaically, the drainage basin, a complex of integrated drainage areas ranging from tiny declivities with their rivulets to the deeply sculpted passage of the big river itself.

"Watershed" is an odd word, though, if you stop to think about it. What does that system of valleys have to do with a shed? "Shed," it turns out, comes from an Old English word meaning division as in "Distinction, discrimination, separation (of one thing from another)," as the *Oxford English Dictionary* tells us. When "watershed" first came into common English use about 1800, probably on analogy with the German *wasserscheide,* it referred to "the line separating the waters flowing into different rivers or river basins; a narrow, elevated tract of ground between two drainage areas." This meaning persists but is perhaps more often found as a metaphor to characterize an event that significantly divides one set of circumstances from another. Soon after, the word began to refer as well to "the slope down which the water flows from a water-parting." And eventually, in the late nineteenth century, it acquired the most familiar meaning we find today: "The whole gathering ground of a river system."

From such a perspective, the Winooski Valley is a region of daunting immensity, but as river valleys go it is of modest size in keeping with the modest size of the state whose perennial rivers and streams total 7,099 miles, whose lakes total 228,915 acres (not including Lake Champlain, shared with New York).[1] The Winooski River is some 90 miles long, its major tributaries about 108, and who knows how many thousands you would find if you counted all the trickles that lace the hills like capillaries in the body. It grieves me to report that if mere length were the only criterion in comparing those rivers entirely within the state, the Winooski is surpassed by Otter Creek, which flows 100 miles south to north from the Valley of Vermont across the Champlain Lowlands to the lake and adds another 150 miles of tributaries. When it comes to the size of the watershed, however, the Winooski asserts its natural superiority. It is the largest of all those in Vermont, New York, and Quebec that feed Lake Champlain, with about 1,080 square miles or some 11 percent of the state, compared to the Otter's 936 square miles. (By way of comparison, note that both streams must bow to Mark Twain's Mississippi, 2,350 miles long, with the fourth-largest watershed in the world, some 41 percent of the area of the forty-eight mainland states.)

To give the waters of the Winooski and the other rivers their passage to the sea, let it be said that Lake Champlain extends north into Canada and empties through the Richelieu River into the St. Lawrence and from there into the Atlantic Ocean. Also eventually reaching the St. Lawrence are the waters in the northeast corner of the state that flow into Memphremagog, a lake that straddles the Vermont-Quebec border. Streams in the far southwest corner of the state make their way to the Hudson River and down to the Atlantic, while the Connecticut River, by a cruel twist of political history actually belonging to New Hampshire, drains the eastern side of Vermont and flows into the Atlantic in Long Island Sound.

Size of any sort—length, volume, depth, height, extension—has never been an enduring source of fame for Vermont, although from time to time the state has produced impressive figures for the production of lumber, wool, milk, maple syrup, granite, marble, and a few minerals. You do not scan Vermont with a telescope; here are no vast seas or boundless prairies or sprawling cities. Of the six New England states, however, Vermont has the greatest variety of landforms.[2] In Vermont you look close up, for its terrain is exceedingly mixed and convoluted, and major changes in landscape and human habitation occur within a few miles of one another. In Vermont's tight geography and geology, you move quickly from one landform region to another—some writers call them physiographic or biophysical regions—all sharing certain broad characteristics but with distinctive geological features, plant and animal habitats, and human activities as well. From the Winooski Valley it is only an hour and a half by car deep into the Valley of Vermont to the south where marble was quarried for buildings in far-off cities. And even less time takes you east along the Winooski Valley through the heart of the Green Mountains to the world-class granite quarries of Barre. Another hour or so and you descend the Vermont Piedmont (the eastern foothills of the Green Mountains) to the Connecticut River. Not that these regions are bounded by the political boundaries of the state, for all extend beyond its borders.

Add to the variety of landform regions the array of specialized pursuits and experience in my valley alone and it is small wonder that I found myself drawn to a cadre of skilled people who shared with me, spoken or not, a love of the valley, and who provided what I could not—a fund of knowledge that added complexity to simplicity, depth to surface. As I made their acquaintance and drew on their experience, generously offered, I discovered my true subject. It was still the river, still the valley, but river and valley as seen through their eyes as well as my own. The act of disciplined perception rose in importance compared to the things seen. There are many interesting animals in the woods but none as fascinating as the hunter, the biologist, the reader of animal signs. This elevation of the perceiver did not arise from any skepticism as to the reality and deep interest of the things perceived—the world is not reducible to the sum of our perceptions of it—although whether and in what sense we can know the whole lies beyond the scope of this book and beyond my scope as well. We can know the parts in their particularity and, to a considerable extent, in their connectedness. And in the Winooski Valley, as in any settled region, the human and natural realms are inextricably interwoven.

I have spread my net wide, but much has passed through or around. The sheer scope of natural phenomena and human engagement required a severe selection. The criteria for selection were often personal—what piqued my interest—and therefore not explainable by any rational scheme. Certain major economic activities such as those historic granite quarries of Barre and the computer behemoth IBM in Essex Junction receive scant mention. The river forms the northern boundary of Burlington, the state's largest city, but it is Winooski, the city across the river, that receives extended treatment. Management-intensive grazing fills more pages than the dominant conventional agriculture. It should be obvious that I have not written a guide to the region.

This book belongs first to my sources, to the people who let me accompany them wherever they were led by their passionate pursuits, their professional responsibilities, their intellectual curiosity. I owe them more than formal acknowledgment can convey.

2

Headwaters

Water is the blood of the land—always in motion, from the rain to the mountaintops, through the forests and plains to the sea, and so to the clouds again.

—ALICE OUTWATER, *Water: A Natural History*

Coits is kind of an insignificant pond, as I recall. —HILTON ABBOTT

As I drive north, my eyes take in the passing trees, the roadside flowers, the road cuts with their folded layers of colored rocks, the rocks that my geologist friend Chuck Ratté can name as I name the members of my family. I can describe the appearance of these rocks, but if I don't know their names, do I really know them? If I knew only their names and had not looked closely at them, hadn't touched them, would I know them? Names are guideposts, links in a family of meanings. When Adam said "Lion" and "Tiger," did he also say *"Felidae"?* My journey in part is a search for names. But today the journey is physical, my first trek in the headwaters.

Where does a river start? Where does the water come from? Like the ever-changing crust of the earth, but on a much faster clock, water is always in motion. The prime movers are the sun and gravity, but transpiration by plants (water given off by leaves) and respiration by animals play their parts too. The sun, it is estimated, evaporates 80,000 cubic miles of water from the oceans each year and still more from streams and ponds and lakes.[1] Gathered as clouds, the moisture eventually falls and, if on land, either is taken up by plants and animals or is drawn downhill through various channels, surface and subsurface, until it eventually makes its way back to the sea or is evaporated en route.

The upper limit of a river system's headwaters is easy to define; it is that line beyond which water flows into a different watershed. The lower limit is hazier, marked by no clear boundary. When I reach Marshfield and Cabot, however, I am clearly at the headwaters of the Winooski. Although a century or less ago, the landscape might have been 75 to 80 percent open—the pastures and meadows of Vermont's ever-vanishing farms—it is now about 80 percent forested, with important differences in the way water passes over and through it.[2] As Alice Outwater tells us, forest trees "act like giant water pumps, directing water back to the atmosphere," as much as 170 gallons per day from a big oak.[3] Their roots affect the soil mechanically, forging many miles of tiny passages. Thanks to these channels and the corridors made by the many organisms that burrow there, "between 50 and 60 percent of the total volume of healthy forest soil is space." The tree canopy softens the impact of falling raindrops, and the spongy soil absorbs them, passing them on slowly to springs and streams. By contrast, open land takes the full force of the rain, absorbs less of it, and passes the remainder more quickly to the waterways, which may flood and erode their banks more violently than a forest stream. There is less water for underground storage, the water table may drop, springs may go dry. Vermont's catastrophic 1927 flood, as Harold Meeks writes, was made more severe by the extent of open farmland.[4]

From the grand perspective of the water cycle, rivers have no sources; they are merely stages in a global succession. So, too, in the broad sweep of Earth's history, our home territories are not fixed places; they are geologic transients. But people see their native landscapes as secure and everlasting, and I look at the Winooski River and ask where it starts. Some would say that even on this tiny arc of the cycle, in the headwaters, it is impossible to pinpoint where the river starts, for it is formed by an intricate web of streams and rivulets, of beaver dams, of seeping springs and oozing marshes. But this doesn't satisfy. Even at the beginning, there must be a main stem to which the others contribute, or so I want to believe.

What do I find when I look at the maps of the region? Several streams—thin, blue lines—compete, angling in from various compass points like the spokes of a wheel centering on a big hub in Marshfield and Cabot. Some are named; more are not. Townspeople and written sources narrow the field to three contenders: Molly's Brook, the convergence of two streams at the north end of Cabot village, and the outlet of Coits Pond. With my good eyes and hiking legs, and the support of the *Canoe and Natural History Guide to the Winooski River Valley,* I settle the matter to my satisfaction. Coits Pond it is, not because of distance—other streams start farther up—but because of channel width and volume. If you hike the headwaters, it is apparent that the other streams feed the one issuing from Coits Pond, even though in dry spells very little water issues from the pond itself. (I reluctantly concede a certain wisdom to those who say that it makes more sense to speak of multiple sources.)

I drive to the pond, just past an old red camp. Two women walking by say they work in a house at the top of the hill for a company that imports baby clothes. You can't search for the source of a river without faint echoes of the centuries-long search for the source of the Nile, of Julius

Caesar and Cleopatra, of Sir Richard Burton and John Speke. Surely all such ventures must lead one into mystery, into contact with something rich, savage, remote. I drive to the source and find baby clothes.

Strange as it may seem, however, the immediate neighborhood of the pond today more closely resembles its primitive state than it did three or four generations ago. A Cabot old-timer said that his father and grandfather had a woodworking shop on the road and a sawmill on the brook that issued from the pond: "We called it 'Mud Pond.'" As for the shop, "There's a small building there, on the right just before you get to the pond, a red camp. That was what was left of the shop."[5] From pond to Lake Champlain, there's not a foot of the stream bed, not a ripple of water, that has not been altered by man. Human beings have walked the valley for thousands of years, for the past two centuries with a heavy tread.

Snuggled into the triangular north corner of the town of Cabot at 1,325 feet above sea level and said to be named for an early surveyor of the state, Coits is a 35- to 40-acre shallow pond, shaped like an upside-down teardrop. Evergreen trees surround the pond, then give way to mixed conifers and hardwoods as the hills rise. Dominating the basin are four high knobs ranging from 1,534 to 1,751 feet. Various kinds of water plants thrive here, adding their organic matter to the other deposits accumulating on the bottom until at some far date the pond may become a marsh, a swamp, a part of the forest. Ponds are mortal.

I see a pretty pond, nestled in the wooded hills. Chuck Ratté sees what I see, with eyes just as responsive to color, light, and shadow, but he also sees a glacial depression scoured out of the soft bedrock of phyllite and crystalline limestone (a "dirty marble," he says). The glacier not only scoured the bedrock, it covered the region with a fairly thick deposit of glacial till, first a compact layer of scraped material pressed under the advancing ice, then a looser layer of ablation till, the debris dropped as the melting ice retreated. Visiting the site with me later, Chuck points out two obvious examples of the glacier's work: two large boulders of very different origins, glacial erratics picked up somewhere and deposited here side by side. He describes their composition with the minute, technical particularity that to my ears might be poetry in a foreign language only sketchily understood. One of them, the light-colored boulder, is an igneous rock with "beautiful crystals of orthoclase feldspar with good cleavage, smooth surfaces that reflect the light," this in a finer-grained matrix or ground mass with specks of a darker mineral, probably a combination of black mica (called biotite) and a needley black hornblende, accompanied by the glassy sparkle of quartz. I understand a little; I admire more.

The road dams up the pond. Where then, precisely, is the river born? Strictly speaking, it cannot be the pond, for a pond is not a river; it lacks the form. I am forced by this admittedly eccentric logic to look to an exceedingly humble spot, to the iron culvert under the road, the birth canal that channels the overflow of the pond into the shape of a stream and spills it into the forest. This is our newborn river, fresh and moist as it ought to be, tumbling into a life that will not end until it merges its being with the great lake to the west. Forever being born and forever dying, apparently it never dies. To our time-bound eyes it never will, and yet some day it will be reconfigured, convulsed, perhaps even eliminated along with the valley itself in the earth's epic mutations.

It is high noon and mid-June when I step off the road into a morass of tangled trees, giant boulders, swamp, and old beaver meadows with their dead tree trunks silvery in the sun. Here the stream meanders a bit, then drops, cuts into bedrock, forms dark pools, then levels out again in a vast swampy area dense with cedars. After an hour or so when I stop to drink and rest, hundreds of insects cluster around my face: mosquitoes, black flies, and other vampiric creatures I can't identify. Some crawl under my glasses and into my eyes; others go into my ears. My hat band is bright red with blood.

I am not a stealthy Indian. Although a partridge flushes (ruffed grouse, we have been taught to say), a frog jumps, a blue jay flashes in the sun, my heavy plodding silences much of the life of the forest. The beavers have abandoned this stretch of the stream, and the place is curiously dead, like an abandoned city swept of its inhabitants by disease. For all I can see, I could be in the remotest forest of North America, but once, listening carefully, I hear the distant throb of an engine. I walk through hundreds of small white flowers, their fragile loveliness incongruous in the rough gloom of towering trees and mossy boulders. I can't avoid stepping on them.

After two and a half hours, I come upon an old road and experience disappointment and relief. I wanted the upper reaches of my river to be wild as they may have been before the white man came. I also wanted not to be lost in the hills, unsure of whether I was following the right channel.

There is no danger now of getting lost. I see a pick-up and a car off to one side, and I wonder what they are doing here. For a moment I wish I had brought a revolver, as I had thought of doing, and I touch the hilt of the Buck knife riding at my belt. I smile at myself, but I remain alert. I come to a pile of old tires, a collapsed metal building, a heap of broken furniture, a scene of abandonment more complete than the beaver meadow, and with squalor added. When the earth has cooled and man has rejoined the cosmos in unarticulated atoms, his tires will remain as testimony to his presence here.

I come out on the town road that took me to the pond. I sit on a bank at the roadside and eat a peanut butter sandwich. I lean against my pack, daisies, red clover, and purple vetch blooming at my feet, and milkweed coming along. The sun blazes down, the hay stands tall and still in a field across the road, cylindrical heads of timothy fully formed. I feel as if I had completed a long and perilous trek. The topo map tells me that if a crow had followed the stream, he would have flown just under two miles; he would not have carried a pack or negotiated ravines and fallen trees. All my years in the woods as a boy are distilled in memories of smooth oak forests, gray squirrels in their branches, of maple groves surrounded by pasture, of old hillside orchards with deer emerging at dusk. I had forgotten the brambles, the impenetrable thickets, the hillsides jumbled and scored as if flung down by an angry god.

Above the road, the stream had passed from rocks and rills to alder swamp, from mountain brook to valley brook, for I am now on a valley floor although a high and narrow one. As I make my way along the edge of the hayfield, it dawns upon me that I have picked the most difficult time of year for hiking, for all wild vegetation is at its most luxuriant and the hay hasn't been cut. I stay near the edge of the field, trampling as little of the hay as possible. I have the vague sense that I am trespassing, and I prepare speeches of explanation and apology.

In most stretches as I approach the village, the stream varies from 10 to 20 feet in width. I fished too much as a boy not to observe the riffles, pools, eddies, undercut banks, and sunken logs. I notice them without trying as an old soldier will take in slope, crest, and pass even when the guns are still.

I eat a carrot, take a drink, and become aware of a deep thirst that calls for a special quenching. Soon I am sitting on the village common, my back against a tree, drinking a quart of orange juice from the Cabot Village Store. My upper lip sports a big lump, a bug bite, and the back of my left hand is swollen. The bell of the Cabot United Church rings four o'clock.

On the common in front of the church is a granite obelisk: "To the Memory of Cabot Soldiers Who Fell in the Great Rebellion of 1861–1865 'Dulce et decorum est pro patria mori'." Thirty-seven men departed this mountain hamlet some 140 years ago to lose their lives on southern soil. In the sun I can read their names. On the shady sides I must trace the letters with my finger, for nature is slowly obliterating this public testimonial just as it absorbed their bodies. I see that the Perry clan gave three men to the cause, Privates Adolphus, Charles, and Jewett, and who knows how many kinsmen of different surnames died in this hill-country town where kinship was thick. They died in a war that put the infant nation on the map of modern warfare, a world-class conflict of principles, men, and arms.

All English schoolchildren once knew the Latin line cited in patriotic devotion on this memorial—"It is sweet and fitting to die for one's country"—and some knew as well "Dulce Et Decorum Est," the poem of the

Englishman Wilfred Owen, who himself fell in battle a week short of the Armistice of World War I, and who had subverted the theme with heavy irony, saying that if the reader could have heard, in a poison gas attack, the blood bubbling in "froth-corrupted lungs,"

> My friend, you would not tell with such high zest
> To children ardent for some desperate glory,
> The old Lie: Dulce et decorum est
> Pro patria mori.

I think of these fallen men and then of Lincoln, a gentle man but steely commander with heart and mind not inferior to Owen's, who knew as much of blood as Owen did, and who in the great crisis saw fit to keep alive "this mighty scourge of war" until more than 600,000 men, North and South, had fallen, a price he was willing to make his country pay in defense of union.[6]

Below the village, I notice that I am beginning to call the stream a river. It is now cutting its way through the glacial till of the valley floor, with wide terraces visible. Eroded banks, sand bars, and developing bends begin to appear, mimicking in miniature the action of bigger streams. I come upon eight half-grown ducks. They paddle furiously and when I get too close they fly, but only a few feet above the water. Sandpipers hop here and there on the river bank, and one flies into a squawking frenzy. I look for her nest but see nothing.

Although the land is virtually level here, walking is still difficult. Some of the ferns and grasses are taller than I am, and there are thistles up to my thigh sharp enough to pierce my khaki pants, the tangle so thickly intertwined I can't see where to place my feet. I spread an opening with my hands, thrust my foot forward, and hope for a safe landing. Once I trip on a hummock and fall, rolling forward on my left shoulder. I never won a foot race in my life, but I have the endurance of a turtle. I can walk for miles, but after a point the muscles and nerves move out of blind habit with no sensitivity to the nuances of terrain. This is the stage at which skiers break their legs, and I do not want to abort my trip with a broken ankle or an abdominal puncture from a sharp stick beneath a fall.

I come out of the woods at what used to be a sawmill. All is quiet and as I round the mill I find a house and a great array of stones neatly arranged in rows inside the mill and out. No one appears to be home, so I inspect the collection, which a sign identifies as a Stone Museum. Another sign says,

> Come See Museum
> Welcome To
> Please Feel to Look
> at Stones. Please

Stay out of Building
Not Safe No Insurance
Gone After More Stones
    Stone Tree
Custer Nephew Curator

One stone resembles a hen. Another, about 24 feet long, makes me wonder how it was moved. I read later that the collector is Michael Ducharme, known as Stone Tree, who was called to collect by a spirit, who is guided to interesting stones by crows and frogs, and who believes that some of the stones are the skulls of aliens for whom Earth was once a penal colony.[7]

A farmer in Lower Cabot, Chris DeNeergaard, gives me permission to sleep on his land. At 7:15, I stop at a grass-covered outcropping of ledge. Beyond the river, across the valley to the west, are Route 215 (the Marshfield-Cabot road), and DeNeergaard's house and barns. Meadows and pastures surround me on all sides. About fifty feet away is an old farm dump, pre-dating DeNeergaard's ownership, with cans, piping, the remains of a barn cleaner, and what may have been a manure spreader. Earlier I saw an old hay loader, rusting away on its iron wheels, the make still visible—New Idea.

I spread my tarp, blow up my air mattress, open my blanket, apply more bug spray, and eat some yogurt and a peanut butter sandwich. As I lie down, one section of the air mattress loses air; I roll over and another flatulates to nothing, like a whoopee cushion. The sun is almost down. Three deer appear in the pasture to the south and slowly feed their way toward me. Their tails twitch constantly, and between mouthfuls they raise their heads and scan the fields. My first real deer rifle comes to mind, a Winchester .32 Special, a Model 94 carbine, purchased at Huntley's Hardware in Essex Junction with money earned delivering papers and mowing lawns. The deer spot me and run into the woods, but they are replaced by seven curious heifers that look at me hard, as alert as wild animals. The sun is down and I feel the cold. I leave my clothes on, tuck a shoe for a pillow under the flat mattress, and stare at the sky.

I try to think about what my eyes take in, not what I know. The sky appears as a dome as ancient men saw it, hunters gathered by their fire or shepherds tending their flocks. My world is much more secure than theirs; if I am just a bit lucky I will live longer and know less of physical pain. But except for fleeting moments like this one, I will never see the sky as a great canopy or as a realm of the gods, guiding my destiny and that of all people. For me the sky is an illusion, although a beautiful one, an optical trick masking inhuman immensities.

A cloud passes overhead, inky gray at the dense center, pink at the edges. It looks like an indistinct animal, holding its baby in its mouth. The sky fades from blue to gray and more stars appear, isolated ones here and there, and then great patterns, most of which in my ignorance I do not recognize. I see the Big Dipper. When I remove my glasses, the stars grow huge and blurred, splotches of light. Darkness gathers. I put my knife within easy reach and close my eyes. The bugs, a nuisance earlier, have disappeared, and I am left with the sounds to the west: trucks rumbling on the road, a dog barking, a cow bellowing. I change positions; there is no hollow for my hip. I am not quite warm enough. Now and then through the night, I sleep.

3

Growing Up at the Headwaters

. . . as we look back at the last hundred years or so, we can't help but feel a sense of awe at the magnitude of change which has occurred.

—CALEB PITKIN, *Marshfield, Vermont: A Photographic Album 1860–1930*

First are the sounds: song birds, crows, the farm dog, cars on the highway. Then I notice the dew, for everything is wet, from blankets and pack to every blade of grass where I am about to walk. On through hay and thickets, soaked from the chest down, in an hour I reach a farm road, cross the river, and come out on the highway at Amanda's Greenhouse, a thriving small business on the remains of the one of the valley farms. Later I will hear much about life in the valley a generation ago from Amanda and her companion and nursery partner Danny Siceley. I trudge along the road back to Coits Pond, initiating the pattern that will take me to the lake: each leg of the journey I cover twice, hiking or paddling downstream, then walking by road back to my car. At first I thought of arranging transportation—a bike?—then decided to make the hike back part of the game, just to say I'd done it. (Later on, I will hike and canoe the entire river once more, camping out along the way.) This hike is seven miles. I arrive muddy, scratched, bruised, bug-bitten, and drenched with sweat. And very happy. I say to myself that this is what our bodies are made for.

But our hearts are made for human contact. Not so many generations ago, the tight geography of these upland valleys determined a social life equally tight, a dense interconnectedness. Only one generation ago, Hilton Abbott grew up in a social world more closely tied to place than our world of easy travel and instant communication. The life of one boy offers a glimpse of the human ecology of the valley, insight into the ambiguities of a close-knit world. I speak from having in some measure shared his experiences. While he was playing, working, and dreaming in the upper valley, I was doing the same along the lower reaches of the river. Like Hilt Abbott, I too left the valley.

I was telling a friend in another state about the river. "Cabot?" she said. "I have a friend who grew up in Cabot." And so it was that I came to meet Hilt Abbott, the man whose path resembles mine—from small-town Vermont to college to many years of teaching—and the one who sees the valley with the double perspective of native and outsider. I saw and heard some of this doubleness at our first meeting. A man of medium stature, lean, vigorous, with gray hair and beard, wearing a denim jacket, this was Hilt Abbott the countryman. A man of well-formed sentences, of wide-ranging references to the teaching of math and physics, of a certain ironic if affectionate detachment from his small-town origins, this was Hilton Abbott the college professor. His story touches on the constraints of upper valley life in his time.

In obstetrical history, Hilt arrived at the end of the home-birth era in 1940: "I was actually born in a hospital, which was pretty high class." We are told that not until the mid-forties did most upper valley women go to Barre City Hospital to have their babies. Speaking of hospital births, a Cabot woman, one of ten children, said, "That would never cross their mind. I guess they thought anybody ever went to the hospital it was to die."[1] After moving around a bit, the family settled in Cabot village when Hilt was in sixth grade.

In the small world of an upper valley town, not starting school in the first grade there could make a difference. He didn't hunt or fish with the other boys because "I didn't have the history of having lived there. I was sort of an outsider; it was difficult to make contacts." And he wasn't good at basketball even though "One of the facts of life is that everyone in that part of Vermont played basketball." Some relief was offered by a new physical education teacher, hired in Hilt's freshman year, who introduced other sports. Unfortunately he also taught algebra, "and he didn't do such a great job of that."

Hilt was quick to add that it was not all the teacher's fault, for the atmosphere was "decidedly non-academic." I've read mixed reports from people who attended school in the first half of the century. Experience varied from school to school and decade to decade. The schools were often social centers that helped bring the community together; sometimes certain mothers would send food to make sure that every child had a decent lunch. It appears that if school meant a few years of basic instruction with no plans for advanced study, it was more or less satisfactory. If it was thought of as preparation for college, some students found it lacking. One woman said that she did not have "a good basic education for college."[2] Many youngsters, especially the boys, never finished high school. A woman who graduated in 1949 said that there were no boys among the nine students in her class, for "In my days, the boys would drop out of school . . . and they would just farm on their dad's farm."[3] Several of the girls became teachers, however.

Hilt was one of the boys who stayed in school, but despite a number of good teachers, "There was no real academic tradition in town. I didn't get a very good education there. Without a cohort of serious students, it's hard, as you know."

Two chance occurrences raised his sights. Thanks to a perceptive lab director at Cabot Creamery where Hilt had a summer job, he went to the University of Vermont instead of one of the less demanding state teachers' colleges. Hilt said, "I can't go to UVM; I'm just this scruffy kid." The director said, "You apply. You go to UVM." Hilt now sees it as "a kind of branch point. Some insignificant thing happens to steer you in one direction and it changes your life."

The other stroke of fortune was the arrival of a new teacher in his senior year. When a student feels a growing hunger for something more even if unarticulated, an alliance may develop with a discerning teacher. For Hilt it was the English teacher Thomas Jones. "The reason that my professional life has been that of an educator," Hilt said, "is that it was the only model I had for what you do if you want to use your mind. Among the "wonderful teachers" in Cabot, Tom Jones was "the biggest anomaly you could imagine. He was a gentleman, he was cultured, he spoke French fluently, and he was black, an African-American from New York. He and his wife had taken their honeymoon in Vermont and decided that they wanted to live there."

Despite the fondest memories, Hilt had never seen Tom Jones again after graduating in 1958. Life has a way of interfering with good intentions. I did some detective work, and it wasn't long until, with me in tow, Hilt was reunited with Tom and Tom's wife Annemarie at the home of their daughter Debbie Doyle in Fitchburg, Massachusetts.

Hilt explained that he had piqued my interest in Tom's story when he told me that his senior English teacher, "a gentleman and a scholar," had made a great difference in the life of a provincial country boy. "I would blush if I could," Tom said. Slight of frame, now in his late eighties but erect and very much in the conversation, Tom had lost none of his quickness. Born and brought up in Albany, New York, he studied Latin, French, Greek, and Hebrew, and eventually earned an M.A. in education, after which he took the job in Cabot. "He can say 'I love you' in eighty-eight languages," said Annemarie, "'I love you' and 'Goodbye,'" to which Debbie replied, "That's my mother's joke."

This new teacher, a cultivated gentleman, brought to Hilt a whiff of a larger world. It was a world toward which he was drawn but lacked the experience to dream about with confidence. And the expanding horizons were not only educational; they were social and political as well. Our conversation inevitably turned to questions of race, for the country was agitated by racial conflict during Tom's time in Cabot, and an intense period of the civil rights movement was about to begin. Tom and Annemarie, however, said that neither in Cabot nor in Winooski, where they later lived, did they ever experience any racial hostility.

Their religion—they are Catholic—might have stirred some animosity, for anti-Catholic feeling, once strong, had not entirely died out. Cabot now has one church, the United Church of Cabot (a merger of Methodist and Congregational). At one time it had five, none Catholic. Marshfield did not have a Catholic Church until 1926. Catholic families traveled to other communities for church, or priests would come to private homes.[4] In the 1920s, Vermont had its own Ku Klux Klan organizations, as did many of the northern states.[5] Then and now one of the whitest states in the Union, Vermont had few blacks for the Klan to harass. Catholics were the target of choice, and Cabot was one of the towns in which they were threatened. Asked about it many years later, one woman said, "They burned a cross, I can remember that. And Dad wouldn't let us out of the house."[6] A more ominous episode was recalled by another woman who said that "The Ku Klux Klan tried to get a hold there in Cabot and of course Johnny Barnett was Catholic and there were a lot of people against Catholics and then they tried to string him up one night . . . Can you imagine in a little town like Cabot? I remember my father and somebody else got him out."[7]

In Hilt's time and earlier, the French-Canadian immigrants were the most obvious Catholic group in the upper valley. Tom remembered that they were sometimes called Swamp Yankees. At school, Tom defended the right of a French family to keep their daughter out of the baccalaureate service, which was to be held in the Congregational Church. The controversy generated considerable heat. Winding up the anecdote, many details later, Tom said, "To make a long story short . . ." to which Annemarie replied, "Too late." To make a long story short, the daughter didn't participate.

It was Hilt's opinion that such prejudice was common throughout Vermont, for "French-Canadians were moving in, invading the territory; they were one culture invading another culture, and there's always friction on the edges." They may have been feared a little too, Hilt said, because they were often successful farmers. I mentioned my father's admiration for those French-Canadian families who through hard work made a go of farming where some Yankee families had given up. Hilt chuckled: "They had the Protestant ethic."

Hilt told us how other experiences contributed to a wider perspective on life: dating the daughter of a Jewish liberal who taught at Goddard College in Plainfield, a decidedly eccentric institution by upper valley standards, and making friends with "more urbane" students at UVM who "were going south and sitting in at the lunch counters. I didn't understand it all yet, but I watched things begin to happen."

Someone produced a Cabot High School yearbook for 1958. It took Hilt back but reminded him that even then he felt distanced from the official earnestness of the school. "If you rest, you rust" was the class motto: "I thought that was pretty embarrassing at the time." There were nine students in the class, two boys and seven girls. I intoned the school's alma mater while Hilt groaned.

Nearby Winooski's Waters
Stands our dear old School,
Where we learned to live together
By the Golden Rule.

We try hard to do our best
For dear Old Cabot High.
We'll leave Cabot with great
Memories, sad to say good-bye.

I had talked about work with Hilt at our first meeting. He never worked on farms as many boys did, but the upper valley was a farming community and farm stories were part of the local lore. When cows were tied in the barn for milking, their manure dropped in a gutter that had to be cleaned out either by hand or by a mechanical cleaner, a long chain with arms that pulled the manure along and dropped it outside. One day an equipment salesman approached Hilt's uncle, a young struggling farmer, and said, "You could use a gutter cleaner. I'd like to sell you a gutter cleaner." And Hilt's uncle said, "Well, by God, I got a good gutter cleaner already, and she makes good doughnuts too." Another story that made the rounds was about a local farmer whose dairy operation was productive but none too clean. One day when the bacteria count in his milk had been high two times in a row, the Cabot Creamery tester looked at the farmer and said, "The next time you come down, would you put the cow shit in one can and the milk in another?"

Hilt's father had been a farmer (not a very successful one), a skilled handyman, and a jack of all trades with whom Hilt shared an interest in technology and process. We talked about the variety of things that were made in small towns before industry became consolidated—rifles, tools, barrels, sleighs, to name a few—and how this fostered highly developed skills and community self-reliance. Sometimes local ingenuity took an unusual twist as in the case of one of Hilt's uncles in East Calais who was a fiddler and a tinsmith, and who earned the nickname Strad for making a fiddle out of tin: "It didn't sound very good, but at least it worked."

Hilt could still describe the equipment and method by which he made butter on his summer jobs at the creamery, and he happened to preside over a major transition in the process, the introduction of an "ingenious machine" from Canada. Instead of a lot of hand operations to produce the packaged product, "You threw the butter in a hopper and it wrapped it in the now traditional quarter pounds," a sharp improvement in efficiency and another step away from the immediately observable, the story of modern technology in a nutshell.

So much for the butter, but what about the buttermilk, a by-product? There was no market for it, so it went untreated down the drain and into the river. "But the real issue for the Winooski River," said Hilt, "was the waste products of the cheese industry. The butter was sort of minor." To make cheddar cheese, you first curdle whole milk and then cook it until you have a small amount of curds and a lot of whey. The whey, drained from the cheddar, went right into the river where it "nourished the slime, molds, algae, whatever it was." In the summer "There would be festoons of slime in the river. It was really very disgusting and nasty," worse than the sewage from homes along the river. And in the late 1940s a second cheese company was founded and added its whey to the stream. One resident of Cabot said that when she first came to town in the early 1950s, "The smell in the upper village was pretty bad, but when you left to go to Lower Cabot you'd roll up the car windows until you'd start up Southwest Hill."[8] In contrast to the "gray slime" she observed, another person reported a more colorful scene, a river that ran "green and purple and when the sun set it was sort of iridescent. When the water was low it was terrible, it was terrible."[9] Even in Hilt's time, however, the plant was beginning to spread the whey on farmers' fields as fertilizer, a practice that continues today.

Hilt recalled a peculiar problem with river pollution. When his aunt and uncle's "great old brick farmhouse" near the Winooski River in Lower Cabot caught fire one day, the volunteer fire department couldn't pump from the river because the creamery waste clogged the strainers, and "The house was lost because the river was so crappy." With a writer's heartless zeal for detail, I said, "That's a great anecdote. That's wonderful. I mean, it's terrible, but for my purposes. . . ." Hilt gave a dry sort of laugh: "We were kind of annoyed."

The more Hilt and I talked, the more links I found with other people I had met along the river. Faulkner didn't exaggerate in his portrayal of complex relationships in the South, and if you could write as many words about the upper Winooski Valley over the past two hundred years, you would find as dense a matrix. The charm of these connections may not be accessible to people whose origins are more cosmopolitan. I will not attempt to persuade. I will say only that I was delighted to find that DeNeergaard's farm in Lower Cabot, where I camped the night of my first hike, was once owned by Ed Gould and that Hilt's grandfather drove horses for him. Or that Hilt's Aunt Hazel married George Carpenter, and that Hilt's father had built a sugarhouse for George's younger brother

Charlie. Charlie had married Barbara Blachly, an Oberlin art history graduate who, while still in high school, bought a small parcel of land in Cabot from a friend for $100.[10] She, quite improbably, has been there ever since, now widowed, still farming organically and milking her dozen Jerseys despite her spinal scoliosis. Her son-in-law Caleb Pitkin was the editor of the Cabot and Marshfield town histories, and it was to Barbara and her daughter Susan that I turned for more information about the region.

The links extend horizontally in space and vertically in time. We place ourselves in both dimensions. Hilt's ancestry in Cabot stretches back in the male line through Mark, Ray, William, and Nathaniel to Amos (1790–1875) who, with his wife Betsey Knight Abbott (1794–1874) were the ancestors of the Abbott, Walker, and Goodrich families now living in Cabot.[11]

And so it goes, filaments of association along which people move, social pathways as real in their way as the beaver channels I found in the marshlands of the lower valley or the ubiquitous trails in the neighborhoods of the white-tailed deer. As real too, I would venture to say, as the pack habits of the wolf or the community character of a hive of bees. While not as obviously necessary for individual survival and sometimes even harmful—such is the wonder of man, his freedom, his capacity for good and evil—no account of a region is complete without its human ecology, and no account of its human ecology is complete if it does not include these social connections. An utterly lonely life is almost unimaginable.

Of course, no account of life is truly complete. Every account—fiction or non-fiction, literary or scientific—simplifies the reality it depicts. This is true not only because of the unbridgeable gulf between the complexity of life and the confinements of narrative and theory but also because speakers speak from a particular cast of mind and disposition of the heart. This may be obvious. I mention it to warn against the illusion of harmony and tranquility this chapter may convey even as it mentions events not in themselves tranquil. The backward glance, the reminiscent mood, essential to the Cabot and Marshfield books, tends to soften the rough edges of life as it was lived.

This life was often difficult. Although emphasizing the positive as local reminiscences of this sort tend to do, the histories reveal the burdens of poverty, illness, disastrous fires, limited education, grueling work, and a general level of physical inconvenience and paucity of goods that would weigh heavily on us today. A listing of deaths in the early years of the twentieth century attributes a number of them to malnutrition and suicide.[12] There are no instruments precise enough to measure these hardships against the compensations of what Barbara Carpenter calls the tribal life that persisted well into the twentieth century. From long association in one place, three generations of a family would know one another with an intimacy less common today. Family loyalty was intense and there were no secrets. The lack of convenient transportation kept people close to home. There were people in Lower Cabot who seldom if ever went to Cabot village. Barbara Carpenter said that someone, when asked if he suffered in the Depression, replied, "No, we had a garden; we had a pig." But Barbara herself saw a broader picture. She said that when she bought her place, "There were abandoned farms everywhere around here . . . the last gasps of the Depression. Cabot itself was down to probably 500 people. Now we're back up to about 1,200."[13]

4

Poking Around, Poking Along

poke . . . To search or look curiously in a desultory manner . . . To live or proceed in a slow or lazy manner; putter. —*The American Heritage Dictionary*

I step into Amanda Legare's kitchen and feel welcome. Yes, I may leave my car behind the barn, but she'll hang the key right here by the kitchen door "in case the barn catches fire."

I resume my hike where I left off two weeks before, making my way through the dense growth. I skirt a thicket of briers and poison ivy intertwined with a pile of old tractor tires. The milkweed is about to blossom; vetch, daisies, and clover are in bloom; the skunk cabbage is turning brown; raspberry bushes are scattered about; and in wet soil I come across a plant that would be at home in the jungle, up to 8 feet tall with huge, coarse leaves, hollow stalk, and round, white flower clusters. It's a gawky plant, no beauty for the home garden, but I think of the energy driving it to such height in a few short weeks. My mind wanders, and then the whole, wet, burgeoning valley floor seems to undulate at my feet as if it were a vast living thing as, in a sense, it is, every bush, shrub, and tree, every blade of grass, every ugly and beautiful thing thrusting itself into the air.

I see my first trout. It looks like a brookie, about 8 inches long, swimming idly, facing the current, waiting for food. The upper river is not a good trout stream—water too warm from unshaded banks, oxygen level too low, and below the first power plant, damage to streambed habitat from the periodic releases of water through the turbine. A school of minnows scatters when I step from the tall grass to the edge of the river. I stir up the red-winged blackbirds as I walk along. They are low, erratic fliers, beating their way awkwardly from perch to perch, but what they lack in grace they make up for in dress, their scarlet epaulettes edged in white, brilliant against glossy black. I sit to rest. It is very hot. I can smell the water. In the long, scent-deadening winter, I had forgotten that in the summer heat even fresh streamwater has a scent.

From Coits Pond to where I am resting, not far from Marshfield village, I saw few obvious uses of the river by humans, although I know that at one time it powered many mills whose remains have disappeared or escaped my notice. It also would have carried waste from every house along its banks and from the two creameries in Cabot (whose discharges led to the stream's slimy festoons). As a feature of the human environment, the river is a stomach and a gut, and this is true no matter how the water is consumed and no matter how it is discharged, whether through direct dumping, septic filtering, or wastewater treatment. All human beings live in a watershed, and they cannot live without using the waters. As a feature of the natural world, the river and its tributaries are the valley's circulatory system, the blood that nourishes, for water sustains all life.

The river is also where interests clash. Although it is useful to imagine the river in its pre-human condition as a measure for monitoring its health, the ultimate context today is not the pristine watershed, the prelapsarian Eden, but the watershed of human presences. The context is defined by glacial till and septic systems, by rainfall and farmers' fields, by snow melt and culverts, by fish populations and planning commissions. It is not only natural; it is economic, social, and in the broadest sense political. In this light, it is useful to spend a moment with Cabot Creamery, for there is more to the story than the iridescent pollution of living memory.[1]

Passing the site on my first hike and later touring the plant, I was struck by the anomaly of a world-class industry, with up to 13 million pounds of cheese aging in its warehouse, located in a tiny Vermont village that you can drive through in less than a minute. Since 1992, Cabot Creamery Cooperative has been owned by Agri-Mark, Inc., a cooperative of about 1,450 New England farmers with headquarters in Methuen, Massachusetts. It markets about 40 percent of the milk produced in New England, a lot of it going to the Cabot line of butter, yogurt, sour cream, and various cheeses including cottage cheese and its famous cheddar, which has garnered a number of national prizes. Agri-Mark's 425 Vermont employees manufacture cheese and other products in Cabot and Middlebury, run an Annex Store in Waterbury, conduct business operations at administrative offices in Montpelier, and extract proteins from whey at a new plant in Middlebury. The proteins are then sold to a Wisconsin company that markets them in the United States and more than forty-six other countries. The spectacular pollution is a thing of the past.

Such is the modern Cabot Creamery, another strand in the web that ties the Winooski Valley to the world beyond. Like the river itself, its origins were humble. The cooperative was preceded in 1893 by a stock company organized to make and sell butter. This was the critical turning point, an even more radical departure than the formation of the cooperative that followed it, for this method of making and selling from one center eventually replaced the previous system whereby each farm churned its own butter and sold it however it could. Hilt Abbott's great-great-grandfather wrote from Lowell, Massachusetts, that he would be home by a certain date "if I get the butter all sold." A Cabot woman remembered how her grandparents used to make and sell their own butter: "They used to make butter all winter and put it—we've still got some of the butter boxes—and they used to put it in these boxes and set it in the cold, then when spring came they'd load all the boxes on the train and take them to Boston and sell them." The new arrangement seemed an improvement to most farmers. One is reported to have said, "One thing about it . . . our wives are getting fat this summer and we are making just as much money as we did in the old way."[2]

After one or more changes of ownership, the plant was sold in 1919 to farmers organized as the Cabot Farmers Cooperative Creamery: "94 Cabot farmers purchased the building that became the original Cabot Creamery plant. It cost the farmers $3,700 all together—$5 per cow—plus one cord of wood each to fuel the boiler."

The history of the organization from then to the present reflects the dominant pattern of the dairy industry over the century, a pattern of consolidation, centralization, and greater attention to environmental concerns. The great majority of dairy farmers have been forced off the land by the need to incorporate radically improved breeding and feeding practices, vastly more costly and sophisticated machinery, and a higher level of management skills. The few who remain produce more milk than ever. In 1989, Bob Davis, long-time top manager of the creamery, said that the nineteen remaining co-op farmers in the town of Cabot itself probably produced ten times the amount of milk produced by the seventy-seven member farmers when he joined the co-op in 1952.[3] A Cabot woman born in 1915 said that there were 16 farms in town in 1990 compared to 150 when she was a little girl.[4] The creamery itself contributed to the consolidation of its part of the industry, buying other creameries or absorbing their producers. It was not immune to the financial shocks of the larger economy or of its own management decisions, however, almost going under during the Great Depression and, much later, merging with Agri-Mark at a time of impending financial disaster. Its efforts to reduce the pollution of the river were in response to new state requirements, part of a general program to clean up the rivers of the state.

There is little in the valley, from milk to river water, that does not lead to someplace else, at least in the mind. The body, however, my body, is making its slow way along the stream, and now I come to a conspicuous human use, a small, red brick hydroelectric plant, Green Mountain Power's Marshfield 6, which takes water through a penstock from Marshfield Dam and discharges it into the Winooski. Also entering the river here is

Molly's Brook, tumbling off the hill in a stretch known as Molly's Falls, a drop of 180 feet over a run of 495.[5] It took the Winooski 6 miles from Coits Pond to drop 433 feet. All the tributaries so far were like the fine strands of a spider web, integral to the whole but individually unremarkable. Here is a tributary with presence and character. At one time, as I have noted, it was considered by some to be the Winooski itself, the trunk stream. It is even now more dramatic than the river with which it is about to merge, but the river absorbs it and flows on.

As I come out of the woods and into a long hayfield, I find dead elms fallen, carcasses of these most graceful of Vermont's trees, once abundant. The Dutch elm disease doomed them, but they are useful in death, for I see big oval holes made by the pileated woodpecker, the state's most spectacular one, black and white, red-crested, as big as a crow at nearly 20 inches long.

Hearing traffic on both sides of me now, I know that I am at the edge of Marshfield village near the fork of Routes 2 and 215. Poking along, I come up from the river into a cemetery across from the fire station. No doubt there are fallen heroes here as in Cabot, but my eyes are drawn to other stones, some expressing confidence in the life everlasting but at least one seeing no farther than the grave. O. and E. M. Smith put their children in the ground with these words:

> Sleep on sweet babes and take thy rest
> God called thee home. He thought it best.

But the family of Mason Woodcock, who died in 1865 at age fifty-five, said farewell this way:

> Another light gone out
> Another home in gloom.
> Husband. Father. Born away
> To the cold and silent tomb.

Although Marshfield is for me mainly a land of the living, I look for the site of a dead man's woodworking shop, for although the building is gone the man has, in a way, continued to live. Arch Cole came to Marshfield as a boy and made his life here. In 1901, he and his mother bought a woodworking shop, where for many years he did custom jobs and made chair stock. He died in 1948 but gained a certain immortality because of the thirty-year record he kept of local events: 1911 to 1941. He called it his log, five volumes of entries but with some notes written on the walls of his shop.[6]

As in the spare chronicles of ancient times, the events speak for themselves without embellishment and are all the more graphic for lack of commentary. Here you find fires, storms, accidents, the influenza epidemic, the progress of hydropower, the price of eggs and beans. Fires were often devastating given the limitations of firefighting equipment and the construction and contents of farm houses and barns. On December 5, 1911, "Harry Martin's barn burned, together with 20 cows, 18 hogs, 7 calves, hens, 175 tons of hay. Loss $8000. Set by spark from gasoline engine. Marshfield's chemical engine used the first time at this fire and made the 5-mile trip from the village in 58 min."

As for weather, snow and cold attract the most frequent comment. On March 19, 1917, Arch Cole noted that "We have 40 in. of snow on the level, some roads impassable with drifts 10 to 20 ft. deep. . . . This winter, 1916–17 there have been 41 days below zero." December 30, the following winter, was the "Coldest day ever remembered. 50 below zero." February 1934 was the "Coldest month for all time. Average 13.9 degrees below zero." The two most dramatic weather-related events, however, were not cold snaps or winter storms but the flood of 1927 and the hurricane of 1938. In the typed excerpts I have seen, the flood is the longest entry, nine lines, and is entirely factual with no explicit expression of emotion—"Big rain. Biggest storm in 60 years. All bridges down the river washed out"—and so on through the number drowned, the cost of damage. Equally tight-lipped and intense is the account of the hurricane that swept across Long Island and New England on September 21, 1938: "Big gale; wind blew at 60 mi. per hr. No lights, wires all down. Lots of barns, hen houses and silos blown down. Rivers flooded. 2,000,000 ft. of lumber down in town. $500,000,000 loss in new England, and 60,000 homeless. Half the sugar and apple orchards gone. $350,000 damage in Vermont." I've read elsewhere that this was the storm that killed close to 700 people on Long Island and in Rhode Island, that flattened 3 billion board feet of timber over a quarter million acres of forest, and that caused more property damage than any other storm up to then anywhere.[7] It was also the storm that my mother feared would blow down the little house in Pittsford, Vermont, where we lived when I was two years old. She was alone; my father was at work. She held me in her arms until it blew over, slashing its way north to the Winooski Valley and beyond.

Cole noted when "Cyrus Sulham, Paul Pitkin, and Will Tibbetts each shot a buck"; when Ozias Pitkin held a sugar party; when Lyle Tucker was killed by a 7-inch falling limb that "fractured his skull, broke both arms, took off nose." On June 7, 1925, he wrote that "Glen Powers and wife and Wavie Sulham's boy buried today; funeral at Congo. Church with Ku Klux Klan service. They were drowned by Lightning Ridge in the Winooski River at East Montpelier." He followed the progress of the new power station at Molly's Falls: water running through the penstock for the first time on June 19, 1926, work started on the new dam on June 21, "water onto the wheel" on December 22, and finally, on January 26,

1927, "Electricity from the new power house for first time. It has been twelve years since power was last generated at Molly's Falls." On February 7, 1937, he "Saw the only airplane to go over this winter," and on March 27, 1936, he "Saw first robin."

In superb third-person deadpan, he recorded that in September 1939 there were "Editorials in Boston Globe and Rutland Herald about a Marshfield, Vt., woodworker who writes significant dates on his shop walls. Also the Saturday Evening Post which says Arch Cole, age 71, has been writing the dates since he began operating his shop 41 years ago." I suppose that even today the things that move us most are the immediate events of family, neighborhood, and work, but with television coverage of wars half a world away, with television program characters we see more often and know better than our neighbors, with instant access to Internet mates anywhere, any time, we are far removed from the pervasive localism of an earlier age.

I move on. Because seeing is more important than going, I try not to think of my advance downstream as progress, but I can't choke back a little surge of triumph when I discover below Marshfield village that I have reached navigable water. Much of the way from now on you could paddle and float and look at the tangled stream banks from the lordly comfort of the canoe, although here and there you would have to get out and walk.

There is no pressure now to get anywhere in particular, so I make my way slowly along the river bank. Stepping onto a sandbar, I hear a startled snort in the bushes and know that I have alarmed a deer. The river is as much a neighborhood to the animals as my village is to me, and much more necessary. Deer tracks are common and unmistakable; others I take to be raccoon, muskrat, squirrel, and skunk. There are bird tracks too. Insects crawl, leaving barely perceptible tracks or none at all. I often find myself following deer paths along the river bank, for they take the easiest route through the ravines and over the knolls, and here or there branch up to higher ground or down to water. Adapted to the terrain, these paths and the routes of smaller creatures exhibit no mechanical regularity, but if we could see them from above as we see roads on a map and if every water course were similarly visible, the valley would stand revealed as a great body of articulated and interwoven parts with vital currents flowing in every vein. And human trails, too—who is to say they do not belong—following as they do the rise and fall and curve of the land, part of the pattern and dense with life.

As I poke along, my energy waning with the day, I look for a place to sleep. I follow a side road and arrive at a farm: two houses, an old barn, two pigs in a pen. I knock at the first house. Such is life in my tight little state of Vermont that this turns out to be the house of a man whose brother I know in my home town of Dummerston. Les Snow and his wife Lori Baker not only give me permission to camp on their land, they invite me to a picnic supper on the lawn. As much as I like peanut butter sandwiches, water, and solitude, I am happy to stretch out on the grass to eat a real meal with Les and Lori, their children Kylee and Jed, their friends in the other house, and their two dogs Jack and Cricket.

I learn that my hosts have formed a little community of part-time farmers, the Wellspring Land Cooperative, dedicated to the wise use of the land and to as much self-sufficiency as circumstances allow. Later a third house is to be built. I am given a tour of the old post-and-beam barn. I see the baby barn swallows in the stable and the goats Lucy and Bart, mother and son, now unhappily separated so that Lori can have the milk for yogurt and soft cheese.

It is nine o'clock and getting dark when I climb the slope behind the barn and find a place to sleep at the edge of a high meadow. A deer snorts in a gully to my left. I lie awake a long time, listening to the humming and buzzing of a million insects and the dull roar of tires from the highway. Sounds travel easily uphill. I sleep off and on, wake up at 2:30. The moon has risen in a curved sky, and I think of the words "under heaven's dome." I can read my watch by moonlight. At 6:20, I crawl out of the bag and into heavy dew. Hundreds of birds are singing and there are more tires on the road. Folding the noisy tarp, I startle a deer feeding on a knoll across the gully. I can see Bart and his mother looking at each other through the fence behind the barn. For breakfast I have a peanut butter sandwich and water. The fog lifts a bit as I walk down the hill, through the barnyard, to the road. From the farm back to Amanda's Greenhouse is about 5 miles. I feel the odd pleasure that we get from closing loops. For the next stretch of the river, I will bring my canoe.

5
Hap's Mill

If you said you wanted a board planed 1 inch thick, it would be planed 1 inch thick.

—DANNY SICELEY

Short, wiry, and strong, with a ready smile, gentle manners, and a quiet sense of humor, Harold (Hap) Hayward looked younger than his years. Only his white hair and a slight hearing loss, caused by his mill and hunting and trap shooting, betrayed his time of life. As we sat at a big table in a room looking over the river in Marshfield, his wife Nakote brought coffee and cake, and I heard the story of his mill. Their house, next to the mill, is as close to the river as you can get and stay dry. No one I met in my travels had both lived and worked as long as Hap in such close proximity to the river.

One of his earliest memories was of the river at its most violent. When his father died, the family split up out of necessity—"We were bounced around"—and Hap was three and one-half when, with his mother and sister, he went to his mother's cousins, an old couple in Middlesex, whose farm soon after took the brunt of the notorious flood of 1927. Described in Klyza and Trumbulak's *The Story of Vermont* as "perhaps the most significant natural event in Vermont's history . . . more destructive than any Vermont flood in post-Columbian history," the '27 Flood killed eighty-five people, fifty-five in the Winooski Valley alone; the little town of Bolton lost twenty-six.[1] The state's property damage was estimated at $25 million, a staggering sum at the time. In the aftermath of the flood, in 1933, Vermont's resistance to outside interference—and help—finally yielded to federal support for flood-control dams in the valley and elsewhere. At the Middlesex farm, "Everybody was scared to death," as the water rose 30 feet, almost to the barn eaves. Despite its potential for violence, however, the river was a necessary partner for some 150 years of the nearly two centuries that a mill has occupied the site of Hap and Nakote's home.

Comfortably retired, a father and grandfather, with many years of productive enterprise behind him, Hap had earned his position as a respected local elder. The Marshfield photo album is dedicated to him in recognition of his enthusiastic help with the book, his "faithful stewardship of the Marshfield Historical Society, his knowledge of and affection for his home town." No small part of his successful life, however, was his mastery of immediate circumstances, his ability to rise above mere coping. Even sedentary people have to come to terms, at least minimally, with the physical world of natural forces such as gravity, leverage, friction, and balance, and with human contrivances such as tools, machines, and buildings. But many remain strangers, never quite at home here. They are surrounded by things they don't understand, from the engine under the hood to the plumbing under the floor, and they either carefully avoid encounters when possible or bump and stumble their way along, baffled and intimidated. On the other hand, some people acquire a special intimacy and facility with the physical world that gives them an ease and belonging the others can only envy; they are the true natives. I suspect that the percentage of aliens, so to speak, is rising because of the technical complexity of the services and instruments at our disposal. Not one web surfer in a million understands how the machinery and system work.

Hap is one of the natives—or was before the information age. He has always been curious about how things are made, from the source of power to the finished product, and he has lived in a time and place where processes could be seen and understood by an inquisitive observer. He told me about the early production and use of electricity in Cabot and Marshfield. One Merton Beckley installed a dynamo at a dam right in Marshfield village in the late 1890s and in 1901 to 1902 formed a company soon known as the Molly's Falls Electric Light Company to build a penstock and generate power at Molly's Falls in South Cabot. Apparently the plant was not physically impressive, for Caleb Pitkin, in his *Marshfield, Vermont,* tells us that one visitor said that "It looked as though Beckley had whittled it out of wood."[2] This is the site that Green Mountain Power later bought, along with other land, and in 1926 to 1927 developed into the system still in use, the Marshfield 6 plant.

Even more open to observation than the early generation of electricity was the direct use of water power to drive the machinery of the mills. In my travels, I saw a number of dams or their remains, slight testimony to this use of the river, but in Hap's childhood there were many more, and before his time a much larger number still. Over the years in Cabot alone, Pitkin writes that there were mills "not only for sawing lumber or grinding grain but for weaving cloth and fulling cloth and carding wool and making starch and tanning hides and making shingles, chair stock, butter boxes, cheese boxes, clothes bars, coffins, caskets, boot-crimps, tennis rackets, bobbins, and even miniature baseball bats."[3] For intimacy of acquaintance with the making of things, only handcrafts could offer more.

Hap as a boy and young man saw the last stages of this means of production, and they fascinated him. He described for the Cabot history how power was transferred from the water-driven turbine through shafts, gears, pulleys, and belts to the various machines in Chub Clark's sawmill in Lower Cabot, an operation Danny Siceley also had observed. Noisy and complicated, the arrangement needed frequent adjustment, but Chub kept his composure—a broken belt elicited only a mild "Aw, shit," as Danny remembered—and always maintained a slow pace, as recalled by Hap: "One thing about Chub, he never hurried, took him forever to put a shoe on. He was slow-speaking too, very slow. A nice fellow."[4]

That Hap is a native, at home in his time and place, was nowhere more evident than in his work, although like many young men fresh out of high school in country towns, he moved from job to job for a number of years before he settled down. He even bought a small farm that came with "One and a half horses. I called a guy up the next morning and he came and took one of them for mink food." In 1952, Hap, Nakote, and a partner bought the sawmill that, along with another woodworking business, was to be Hap's lifework. The next year, Hap and Nakote bought the house just a few yards upstream of the mill. (Hap later bought out his partner.)

"And you've been here ever since."

"I never got enough money to get out of town."

"I doubt that."

The first mill on Hap's site, built in the early nineteenth century, was an up-and-down mill—the blade moved vertically—perhaps an arm and wheel arrangement like the drive wheel of a steam locomotive. (The house came a bit later, sometime between 1812 and the Civil War, and "It slopes a little.") In 1928, then-owner L. A. Farrington installed an S. Morgan Smith hydraulic turbine in a concrete wheel pit at the edge of the river. It is still there and Hap and I went out to look at it.

Moving water contains energy, and the faster it flows, the more it contains, a fundamental fact of river dynamics and hydropower. The object in hydropower is to convert some of the moving, fluid energy (kinetic energy) to a useable mechanical or electrical form. Dams back up the water, creating head or drop and thereby increasing energy. A log dam crossed the river at Hap's mill—it went out in high water many years ago—and all the machines of the mill were run by water power until 1946 or 1947. Hap used water power for only one machine, a blower that blew sawdust into a bin.

In principle, the operation was simple: water from the dam turned the turbine (essentially an enclosed wheel), the turbine turned shafts and belts, and the belts ran the saw and other equipment. In application and practice, it was not so simple. The water came from the dam into the wheel pit through a curved concrete channel called a head race. The curve set up a circular motion to get the water turning in the direction of the turbine action. Water turned the turbine and was then discharged through a draft tube into another channel called the tail race. The tail race was designed to keep the water at a certain height around the lower end of the draft tube, so that the weight of the falling water would create a partial vacuum in the tube. The weight of the water both pushed and pulled. The wheel pit's inlet had wood slats or bars called racks to catch debris carried by the water. Hap had to "rake the racks" now and then, especially in the autumn when "Leaves would plug it tighter than the devil."

The word "wheel" suggests something relatively simple such as a wagon wheel (no doubt a wonder in itself when first conceived), but the turbine is an ingenious, precisely designed machine whose operation Hap pointed out in detail but which I will only sketch. The enclosed wheel, the turbine, has vanes or blades. Around the perimeter are twelve wickets or wicket gates, which are like vertical Venetian blinds that can be opened a little or a lot to change the volume of water that comes in to push the vanes: the more water, the greater the speed of the turbine. This turbine has a vertical shaft, but the shafts in the mill are horizontal, so how does energy move from one to the other? The answer is obvious to engineers, mechanics, and mill operators but not to curious writers. Hap showed me the bevel gears, a solution so clever yet simple that at its invention people must have said, "Why didn't I think of that?" Two 45-degree bevel gears make a 90-degree turn. The first bevel gear has replaceable wooden teeth, which will self-adjust to the iron teeth of the other gear.

Hap has a letter dated March 23, 1928, from the S. Morgan Smith Company to L. A. Farrington (misspelled Farmington) stating the terms of the sale. The 30-inch left-hand water wheel cost $915, two shafts came to $225, and the pair of bevel mortise gears the same, for a total of $1,365. Accompanying the letter is a bulletin with engravings of the turbine and parts, instructions regarding the races and pit, and tables of specifications and performance. The pit must be big enough, for example, to allow a volume of flow sufficient both to run the turbine and to flush out sand and gravel that otherwise would accumulate.[5] A table shows how horsepower, revolutions, and volume of water discharged vary according to the diameter of the wheel and the head of water. At 3 feet of head, Farrington's 30-inch wheel would generate 10.9 horsepower, would turn at ninety-six revolutions per minute, and would discharge 2,375 cubic feet of water per minute. The larger the wheel, the greater the horsepower and the cubic feet but the slower the rotation. The greater the head, the more all these numbers increase but not at the same rate.[6] One's choice of wheel size would depend on the head available and what ratio one wanted between power and speed.

Most of the operations of the mill had been converted to electricity shortly before Hap and his partner bought the place, and they continued to modernize by putting up new buildings. Although the turbine had worked for many years, it had a significant defect: It was too big for the amount of water. If you lost even 1 foot of head, you didn't have enough power for anything except the blower.

Like all technical operations, indeed virtually all human pursuits, the mill has its specialized vocabulary, which first baffles, then informs, the observer. As we inspected the machinery, Hap explained how a log would be rolled on two beams called the rollway to the carriage where you would dog it, that is, secure it with two hooks: the post dog that simply held it and the sawyer's favorite that adjusted the position of the log vis-à-vis the saw blade. The carriage moved the log endwise into the 48-inch circular blade to cut the dimension lumber (2 × 4s and bigger) that Hap preferred to produce. A trimmer saw then cut pieces to length—6 to 16 feet. Hap was known for his good work. "If you said you wanted a board planed 1 inch thick," Danny Sicely said, "it would be planed 1 inch thick."

The board saw was run by a 50-horsepower electric motor that came out of a coal mine in Pennsylvania. In its forty years of service, all Hap has done to it is add about two quarts of oil. His saw rig was ten years old when he bought it from a heel factory in 1954. His 1939 forklift still runs. He showed me the blower, which removed the sawdust; the chipper, which reduced slabs and edgings to quarter-inch bits; and the hammer mill, which pulverized these waste bits into "a stringy mess the consistency of shredded wheat." (The sawdust, waste, and shavings were mixed and sold for animal bedding.) Here is a man who gets along with machines.

In a building across the river, Hap and his partner ran their other business, manufacturing chair rockers and other things to use up the waste: "We were a very small producer, and we produced about a thousand rockers a day. You could hardly believe that they could sell that many rocking chairs. Week in and week out." Hap bought lumber by the log, milled it, and sold about half, mainly softwood, to Alcoa in Messina, New York, where it was used for crates and reels, although some was sold locally. The other half, maple and birch, went to rockers. I asked about the supply of logs for rockers. All was well until about 1965 when the Japanese started buying logs for bowling alleys, and "They got the price up where we couldn't make it." The rocker business closed in 1968.

Japan isn't the only player, Hap said: "Right today there will be ten to twenty trailer loads of logs going by the house on their way to Canada," the Canadians paying more than a small mill can afford. They can do it because "They've got better machines. There's a mill in Sherbrooke [Quebec] where they claim they can saw it faster than you can unload it from a truck with a cherry picker." And then the lumber comes back to the

States to be sold. (Canada also allegedly subsidized its lumber industry, but in recent years, Hap later told me, the United States imposed a duty that helped to equalize things, and there was no longer a parade of lumber trucks headed for the border.)

There are networks everywhere. Looking at the valley requires two lenses: a close-up for the particulars of this discrete environment, a wide-angle for its multiple connections with other times and places. The distinctively local met forces from afar in the life and work of Hap Hayward. Fortunately he had the mind and character not only to hold his own but to enrich the community with his presence.

It was time to say goodbye.

"Are you completely out of the business?"

"Haven't cut lumber for a year. I could. Oh, I sawed a little last summer for somebody. No one to take it over. If I sold it I would have to sell the house. I'm here and I guess I'll stay here."

Hap will stay; I must move on. As I am about to descend deeper into the valley, it is time to explain its origins, and for this I will need my friend Chuck Ratté and his fellow geologists.

6

The Making of a Valley

This is my Father's world . . . —OLD HYMN

Streams shape the land. —GEORGE SPRINGSTON, Geologist

The story of the Winooski is the story of glaciation.
—CHARLES COGBILL, Historical Ecologist

I'll turn for a moment to the other end of the valley. Below the falls and dam where the Winooski River divides the cities of Burlington and Winooski, there is a magic place called the Salmon Hole. One July morning, I followed Chuck Ratté down the steep path on the Burlington side to some rock shelves on which he showed me raindrops, mudcracks, and the ripple marks of waves from half a billion years ago, inorganic fossils preserved when a sudden wash of sediment covered them. Over the ages, layers accumulated and then eroded, revealing to the trained eye a micro-moment in the history of the earth. What Chuck saw, with a fascination undiminished by a lifetime of professional study, was not just the operation of general laws but the particulars thus produced. The particulars were unique, a certain array in a certain setting, a small sample of the valley's innumerable variety not just of preserved raindrops but of all the features of the natural and human environment. The trained observer sees with a double vision: the particulars in their individuality, sometimes beautiful and always interesting, and the general with its inexorable grandeur. Or so one would hope.

We were standing near the shore of a one-time shallow sea, the rocks had a history and a name, and I knew then that it would be through Chuck's eyes and the sources to which he would direct me that I would come to know a little of the deep foundations of my valley. Foundations in several senses of the word: the shifting of the continental plates that molded the deep structure; the glacial, stream, and weathering action that sculpted the surface features; and the collaboration of these forces with chemical and biological action to form the soils on which most life would depend. Exploring the valley, I would move beyond geology in the narrow sense, but from trap line or bobcat ledge, from farm or power plant, even from village and city, it was but a few steps back to the foundations that Chuck Ratté had devoted his life to studying.

And why did he devote his life to it? Tall, handsome, lean, and fit well into his retirement years, he could have been, I tell him, a movie star, the Indiana Jones of geology. At Middlebury College, he was inspired by the enthusiasm of a teacher to major in geology, then earned a master's degree at Dartmouth and a doctorate at the University of Arizona. "Geology," he says, "is better than working for a living." Early on he developed a passion for a "practical knowledge of the planet and its workings," a passion that never abated in a long career as industrial geologist, college teacher, and Geologist of the State of Vermont.

For many generations, the planet and its workings have stirred thoughts and feelings of a different order. My grandmother, a staunch Christian in the Puritan mold, sang with fervor and conviction these lines from an old hymn:

> This is my Father's world,
> And to my listening ears,
> All nature sings, and round me rings
> The music of the spheres.
>
> This is my Father's world;
> I rest me in the thought
> Of rocks and trees,
> Of skies and seas;
> His hands the wonders wrought.

She saw wonders. She was so constituted by nature and experience as to regard the natural world with the wonder that is said to be the beginning of philosophy, but for her it culminated in faith. Modern geologists see wonders too, but where my grandmother saw the hands of God, these scientists, whatever their religious convictions, see the unvarying operation of natural law, the profound significance of which is easy for the layman to overlook today when its truth is so widely, if not universally, held by educated people.

One might fear—many did fear—that the belief in unalterable general laws left no room for divine providence and stripped the earth of all special significance beyond the mechanistic realm of matter and motion. The result for some was the "disenchantment of the world."[1] I found a new enchantment, for geology shares with all modern science a history of discoveries that defy common sense and stagger the imagination.

We can say that we understand the age of the earth, some 4.6 billion years, or the vastly greater immensities of time and space of the known universe, but in fact the disproportion with our human scale is so great that we can only grasp them abstractly; we cannot imagine them. Easier to imagine and now a commonplace in popular literature is the movement of the earth's crust. But think what common experience had to be transcended for the idea to make sense. "Keep your feet on the ground," we say, as if that were a secure place to be—*terra firma*—as indeed it is, most of the time, in the scale of experience on which we live our lives.

The story of the movement of the earth's crust, the tectonic plate theory, has been told so frequently that it needs but the briefest account here.[2] In the early twentieth century, evidence including the jigsaw-puzzle relationship of Africa and South America and the presence of warm-water formations in temperate latitudes such as the coral reef in Lake Champlain led scientists to believe that the continents must have moved.[3] In the fall of 1956, my college geology text noted that mountains were known to be still rising and folding, but the forces that moved them were "mysterious," and "No one knows the ultimate cause of earthquakes."[4]

The text presented as a possibility the convection theory, which has come to prevail. Refined over time, it holds that currents of hot material within the earth rise, spread, and subside, causing the various segments or tectonic plates of the earth's surface to move, to separate and collide, sometimes rising as mountains, sometimes sinking and melting to emerge again as volcanoes. The tension and sudden release as moving plates grind past each other along so-called faults cause earthquakes. And what goes up must come down, thanks to the erosive forces of heating and cooling, ice, streams, wind, waves, and the impact of plants and animals. The cycle thus understood amounts to a general theory of the earth. It was one of the major scientific achievements of the twentieth century.

Chuck Ratté said that to understand the processes that formed the Winooski Valley and determined the human uses to which the region could be put, one had to know something about the rock cycle. Particles eroded from mountains are compacted (lithified) into sedimentary rock layers of varying density depending on the kind of particles, the degree of pressure, and the presence of cementation by minerals: clay becomes shale, sand becomes sandstone, coral skeletons become limestone, to give just a few examples. Then, in the movement of the earth's crust, two plates may collide, squeezing and folding these sedimentary layers, and forcing them up in the mountain-building process geologists call an orogeny. The pressure of the folding produces the denser metamorphic rocks; in our examples, shale becomes slate, sandstone becomes quartzite, limestone becomes marble. Another possibility is that a heavier plate may sink or subduct under a lighter one. The rocks of the subducted plate then melt in the intense heat deep in the earth, becoming a molten material called magma. As magma cools, rapidly or slowly depending on circumstances, it forms various kinds of igneous rocks, including the familiar granite of the Winooski Valley. The magma may erupt to form volcanic mountains, sometimes along zones where plates are pulling apart. And all the while, the forces of erosion are wearing away at all kinds of rocks, producing more sediment for new lithification and, along the way, sand and various soils for the farms and gardens along the Winooski River.

The rock cycle, then, is part of the larger circularity of mountain building and destruction. As Chuck and I scanned the mountains and valleys on our field trips, much of the rock cycle was in progress all around us, but so great is the disproportion between our human time scale and the earth's, we could not see it directly. Only in volcanic eruptions, which Vermont is currently spared, or in erosion, which Vermont now has in abundance, can a human being actually see the rock cycle in action.

But what about the epigraphs at the head of this chapter: "Streams shape the land" and "The story of the Winooski is the story of glaciation"? A contradiction? Not really, but before we can reconcile these remarks we have to go behind and below both streams and glaciers; we need to visit, if only briefly, the realm of deep geology. Before the sculptor can carve and polish, someone must bring him his stone; rivers and glaciers need something on which to work. Chuck was my guide as I traveled backward in time and downward in space, to ancient epochs and deep convulsions.[5] What we see is part of who we are. The boy who felt the warm sand between his toes still lived within me, but now I tried to see beyond the context of sand bank and oak woods. It was a long way I had to look, for the ancestral line of the Appalachians—of which the Green Mountains and the Winooski River are a part—stretches back over a billion years.

Roughing Out the Stone

Once perhaps as high as the Rockies, the Appalachian Mountains underwent successive waves of formation and erosion until today the highest peak is North Carolina's Mount Mitchell at 6,684 feet. In the current arrangement, New Hampshire's White Mountains are geologically distinct from the Appalachians, which is fortunate because it gives Vermont's Mount Mansfield second place among the Appalachians at 4,395 feet. Speaking of all of eastern North America, however, Mansfield is topped by some Maine and New Hampshire mountains, including the highest after Mitchell, New Hampshire's Mount Washington at 6,288 feet.

The very beginning of the story would be the formation of the earth, but the farther back we go, the scantier the evidence, and eventually it disappears. My time travel started more than a billion years ago with a North America not shaped as now or in its present location but much farther south in tropical waters, and with what is now Vermont then a flat coastal area. The shifting continental plates brought about three orogenies or mountain-building events—the Grenville, the Taconic, and the Acadian—that helped shape the Winooski Valley, here briefly mentioned but described in more detail in other texts such as Klyza and Trombulak's *The Story of Vermont*.[6] (A fourth orogeny, the Allegheney, affected only the southern Appalachians.)

About 1.1 billion years ago, two tectonic plates along the east coast of North America smashed into each other, pushing up the entire area into high mountains, the so-called Grenville Orogeny. It appears likely that the joined plates were just part of a worldwide convergence, the Grenville Supercontinent, a predecessor to Pangaea, another worldwide landmass. By 740 million years ago, the mountains of the Grenville Orogeny had worn away, but they left embedded rocks that through later mountain-building events would reappear in present mountains including the central Green Mountain range and the Adirondacks. Such was the first ancestor of the Green Mountains of which we have good evidence.

Eventually plates shifted and the supercontinent began to break up. A shallow ocean formed, and into it the eastern edge of the North American land mass deposited its eroded sediments. These sediments eventually became the limestones and shales we find in the Champlain Valley and the Valley of Vermont.

If it was a relief to me to find this slight visible connection with the landscape that I know, the next chapter in our geologic history brought the distinct pleasure of seeing our child's features in the photo of a grandfather. Orogenies are hardly personal events, but I have developed a certain affection for the Taconic. Some 450 million years ago, after various convolutions, the collision of two plates produced the Taconic Orogeny, mountains that have not entirely disappeared, for among their remains are Vermont's Green Mountains and the Taconic Mountains of Vermont, Massachusetts, Connecticut, and New York.[7]

The third orogeny, the Acadian, helped to prepare for the Winooski Valley by forming the geologic substructure of the headwaters. About 400 million years ago, magma pushed up in vast granite formations, including Signal, Burnt, Butterfield, and Knox mountains, some of which Chuck and I could see as we looked east from Montpelier. Some are volcanic in appearance, but they are in fact mainly granitic monoliths that resisted the more extensive erosion of their surroundings. Also Acadian are the huge granite deposits in Barre, which, if quarried at their present rate, are estimated to last a thousand years. I came to see that although I had always thought of all the hills and mountains from the headwaters to the beginning of the Champlain Lowlands as part of the Green Mountains, historically there were two separate orogenies, the Taconic for the central ridges, the Acadian for the territory of the headwaters.

But to return for a moment to the Taconic Orogeny: Not only did it provide the gift of the original material for much of my valley, it was probably the geologic drama behind the event that consecrated, as it were, my entire enterprise. Chuck and I had made our way to a site perfectly familiar to geologists but new and a place of enchantment to me. If the science of geology could be imagined as having attendant spirits, surely they would hover about the Champlain overthrust fault where it lies exposed in Burlington.[8] Running north-south in Lake Champlain and one of several parallel faults in the region, it comes into view at Lone Rock Point where glacial and lake erosion has revealed two layers of rock made of different material, the older Cambrian dolomite above the younger Ordovician shale (the one some 560 million years old, the other 430 million). Earlier I had grasped the principle of superposition—older layers of sediment were laid down first and are therefore on the bottom. That here the older stratum should lie on top was a striking demonstration of the movement of the planet's broken crust. Ages ago, the enormous force of the Taconic collision had thrust the older layer over the younger.

Those shadowy crustal plates and that abstract process of shifting and colliding—nearly everywhere masked by the blanket of soil, grass, and trees, and further hidden by the apparent stability of Mother Earth and by the selective vision of my love of familiar landscapes—those obscure entities of scientific theory here stood revealed to my senses. My feet made a scraping sound on the shale. My fingers grazed the underside of the dolomite where it was rubbed smooth from sliding over the bottom layer. Chuck described the rock, gave the ages. The names and numbers meant little to me at the time, but the division between the darker shale and the lighter dolomite was right before my eyes. It was a revelation. I felt as if I were present at the creation.

With some understanding of the three orogenies, I knew how the Winooski River, the master sculptor, got its material; later I would find out how it did its work. But now I had some elementary questions about the origins and patterns of the Winooski and its tributaries. Chuck said that the little streams in the hills are known as consequent streams; they began to flow simply as a consequence of topography. Combined, they made larger subsequent streams whose paths were determined by the already existing rock structure; they cut their way through weaker zones such as soft rock or crushed areas, forming the valleys of the bigger tributaries of the main river.

The Winooski itself is something of a puzzle. The main faults and folds in the mountains run north-south, but here is a river that appears not to have taken an easy route but to have crossed the mountains more or less east to west, as did the Lamoille and the Missisquoi, sister rivers to the north. If the Winooski had originated after the mountains were formed, you would expect it to flow south or east into the Connecticut River. It is assumed, therefore, that this is an antecedent river, older than the mountains through which it runs. Already flowing in a valley before the orogeny, the river was able to keep pace with the uplift, carving its way down as the mountains rose up. Since it is "out of harmony with the rock structure," as one writer puts it, the Winooski is also called a superimposed river.[9] Chuck likes to point out that to leave the story there, however, is to present an explanation that doesn't explain. How did the river keep pace with the uplift? The river may have been out of harmony with the general rock structure, but no stream can run uphill or erode hard rock faster than soft. It follows that the Winooski had to be also a subsequent stream, following some east-west line of weakness in the rock formations; space photography, Chuck says, shows a definite east-west lineation beyond the Winooski through New Hampshire. What the weakness was isn't clear; further investigation is needed. Perhaps the rocks moved just enough to create a crushed zone without disturbing the basic north-south alignment of rock formations, which are the same on both sides of the river. My Winooski, the river I knew mainly for its beauty and its fishing, must now assume the dignity and bear the burden of a weightier designation: It is now an antecedent, superimposed, subsequent stream.

It must endure two more classifications before the valley can receive its fine modeling. We return to circularity, for rivers, like living beings and mountains, have a life cycle, and with their tributaries they form patterns in the land.

I was saddened to read in Meeks's *Vermont Land and Resources* that "The Winooski is a tired river."[10] In the fifty miles from Montpelier to Burlington, "Its gradient is only eight feet per mile and it does not have the energy to move very much of the larger material delivered to it." Although tired, the Winooski is considered a mature river, not an old one. Age is not a matter of years; if it were, the Winooski would be a veritable Methuselah. It is the degree to which the river has developed the valley. A young river has a steep gradient and a narrow valley; its energy is expended mainly in deepening the valley. A mature river has a lower gradient; its energy is expended mainly in widening the valley. An old river has a valley floor much wider than its meanders and therefore has stopped widening the valley and does very little downcutting. Although mainly mature, the Winooski has youthful passages such as its beginnings below Coits Pond and its V-shaped gaps at the dams in Bolton, Essex, and Winooski. In extreme old age, a so-called peneplain appears ("nearly a plain"), but often uplift interrupts the cycle, rejuvenating the streams. In fact, I've read that although peneplains have existed in the past, none are known anywhere in the world today.[11]

River patterns are described as they appear from above. If you looked down from an airplane on the Winooski Valley with an eye for only the trunk and the major tributaries, you might say that it had a trellis pattern. However, if you took in the lesser streams feeding the tributaries from a variety of angles, you would see a tree-shaped pattern known as dendritic, the whole arrangement resembling a tree with its main trunk, larger and smaller limbs, and many still smaller branches and twigs.

Modeling the Fine Features

As Chuck and I were driving along River Road in Jonesville, he asked me to stop where the bedrock emerged from the valley floor in a rounded hump. We climbed it, "a beautiful, smooth outcrop, glacially polished and with very distinct striations running east-west."[12] The whole ice mass moved northwest-southeast, but local conditions caused lateral movements. And then, as often, I heard the music of a description I could only partly follow, for there we had a muscovite, quartz-rich schist, a product of middle-intensity metamorphism with some crinkles caused by lateral movement while squeezing, and a surface smooth but not as smooth as some because the ice moved at right angles to the foliation.

Streams do shape the land, as we have seen, but the story of the Winooski is also the story of the continental ice sheets that repeatedly scoured the Green Mountains and, at one time or another, covered nearly 30 percent of the planet's land area. The routes and work of the most recent glacier are well known, but the search for the ultimate cause of glaciation has been long and difficult, and even now is beset by uncertainty. The personal and professional adventures of some of the early theorists, among them the Scot James Croll (d. 1890), make for stirring accounts of triumph over great odds.[13] Of Croll we read that "The man who had begun his career as a mechanic in Branchory, operated a tea shop in Elgin, failed as an innkeeper in Blairgowrie, and worked as a janitor in Glasgow, had become a world-renowned figure in science. . . ." Croll's hard-won fame was based on his theory that the alternation of glacial and inter-glacial periods was due to variations in the earth's axis of rotation and in its orbit, both affecting the amount of radiation received from the sun. By the end of the century, Croll's views were widely rejected but were revived and improved upon by Yugoslav mathematician Milutin Milankovitch and others in the twentieth century.

Although not universally accepted as the complete explanation, these astronomical phenomena appear to play a major role in the so-called Ice Age, the cycles of glaciation and melting over the past two to three million years. There were earlier periods of glaciation, however, less well understood because so distant from us—300 million, 600 million, and 2 billion years ago—and for these, other causes have been advanced, including continental drift and changes in the direction of ocean currents. Some scientists now believe that in very early times there may have been periods of world-wide icing when even the oceans froze, then melted when enough heat was trapped by volcanic carbon dioxide in the atmosphere.[14]

Great ice sheets forming some 2 million years ago covered the northern part of North America but did so in a series of major alternating advances and retreats, these in turn further divided into smaller pulsations. The Laurentide, the eastern sheet, advanced and retreated nineteen or twenty times, the most recent thrust being the Wisconsinan, which itself advanced and retreated seven times and after 70,000 years reached its peak only 18,000 to 20,000 years ago and then melted back until by 10,000 years ago it disappeared from Vermont. While a long time in human history, these time spans are but micro-seconds of the geologic clock and are graspable by the imagination, which may be one reason why they stir more than the usual geologic interest among the general public. (I have read that if the span of Earth's history were compressed into one year, Columbus would reach America at 11:57 P.M. on December 31.[15])

Although ice as we see it is brittle, at 200 feet deep or less it becomes plastic and moves with a laminar flow, slowly, like a viscous liquid; it also slides along the ground when tugged by flow or gravity. There is no doubt about the plasticity of the recent glacier over Vermont, for the ice was 1 to 2 miles thick, heavy enough to depress the surface of the earth at the northern end of the state 600 to 800 feet below its present level; in fact, the crust is still rebounding.

Glaciers worked directly and indirectly. They smoothed and rounded bedrock nearly everywhere, plucked and roughened it in a few places, scoured basins for future ponds and lakes, and widened and deepened valleys including no doubt the Winooski.

Glacial ice is very dirty, loaded with rocky debris ranging from the finest sediments called rock flour to enormous boulders. The geologist George Springston says that glaciers are the earth's sandpaper, for the grit they carry can polish if fine enough, scratch if coarser, and gouge if coarser still. That bedrock hump in Jonesville had been rounded, smoothed, and scratched. The most striking visible example of direct glacial action, however, are those hills with gradual north and sharp south slopes known as *roches moutonées* or sheep rocks because of their profile. The ice gripped and plucked out huge chunks of the south face, which were carried away as glacial erratics. A spectacular example, visible from the Winooski Valley, is Camel's Hump, one of Vermont's highest and most easily recognized peaks at 4,083 feet, with its steep leeward cliffs.

The indirect consequences of the glaciers were more momentous. When glaciers retreated, they left behind an immense load of drift. Some of this drift or rocky debris was lodgment till, which was compacted tightly under the weight of the ice but which could be eroded. Some was ablation till, which also could be washed away after being randomly dropped as the glacier melted (still obvious in upland pastures). And some was drift carried away by streams flowing over, under, or alongside the glacial ice.

If carried by running water, the drift was gradually dropped and sorted, large particles to small. If dropped from the end of a melting, stationary glacier, it formed a mound called an end moraine, rare in Vermont, such as New York's Long Island at the farthest advance of the most recent glacier on the East Coast. (The glacier was stationary because the ice melt and flow were in balance.) If drift was deposited between a valley wall and the margin of the melting ice, it formed narrow ridges or mounds known as kame terraces, found in the Winooski Valley and most valleys in Vermont. If it was deposited by streams in still bodies of water, it formed deltas such as the ancient delta we call the Intervale in the lower Winooski Valley near Lake Champlain.

This drift deposited in the Winooski Valley and throughout Vermont has had a profound effect on plant and animal life, including the life of human beings. As Harold Meeks writes, "In short, the recent ice sheets, in scouring the landscape and in covering it with rubble, probably have a greater bearing on what people are doing now with their land than any other natural processes operating throughout geologic history."[16] Worked and reworked over the millennia by the many streams in the valley, these deposits became building sites, sources of sand and gravel, and the good agricultural soil on DeNeergaard's farm in Lower Cabot where I camped the first night of my hike, and on the Whitcomb farms in Essex Junction and North Williston, which I was later to visit.

My imagination is most stirred by another geologic consequence of the last glacier, the succession of freshwater lakes and a salty sea. Writers never tire of mentioning the whale skeleton found in Charlotte in 1849 and now on display at the Perkins Museum of Geology at the University of Vermont. The discovery was remarkable, for this creature, a white whale (also called beluga and other names) was a saltwater animal.[17]

And so are seals. But the early Vermont naturalist Zadock Thompson described two of them, obviously adapted to freshwater, that were killed on the ice of Lake Champlain, one in 1810, the other in 1846.[18] Thompson thought that they had swum up the Saint Lawrence and the Richelieu into the lake, but we now assume that they were survivors of a time when the Champlain Valley was an arm of the Atlantic, the so-called Champlain Sea, with about half the brininess of the main ocean. As such, this sea was part of the evolution of Lake Champlain, a complex drama in which the glacier played a leading role. Before the Champlain Sea, a succession of freshwater glacial lakes in Vermont was formed when the retreating ice, and in southern New England an end moraine, prevented the rivers from flowing into their previous drainage basins; at one time nearly half the state was covered. (The advancing ice also had created lakes many years earlier.) From about 11,000 to 10,000 years ago, the Winooski and other rivers formed lakes in western Vermont when ice blocked their flow into the Champlain Valley. Over the years, geologists have come up with various interpretations and nomenclature. My sketch of the Winooski Valley glacial lakes is a synopsis, minus the supporting evidence, of the views of Frederick D. Larsen, Vermont's premier glacial geologist.[19]

Larsen divided the history of the Winooski glacial lakes into three stages. The first consisted of Lakes Granville, Roxbury, and Williamstown that were formed, respectively, by the north-flowing Mad River, Dog River, and Stevens Branch, all tributaries of the Winooski. The earlier and present watershed divide between the Winooski and the south-flowing headwaters of the White River runs through Granville Notch, Roxbury, and a site south of Williamstown, but as the water of these early lakes built up, they spilled south over the divide at these three thresholds. With a certain poetic license, we might say that the Winooski ran backwards. I feel a personal connection with these events because in the course of my travels I looked for at least one precise point where the waters divided. After several stops along Route 100 in Granville Notch in the headwaters of the Mad River and scaring up a young bull moose who was feeding in a roadside pond, I found myself at the critical point. Facing east on a ridge, I thought that if a persistent and long-lived raindrop were to bounce off my right shoulder, it could make its way to the ocean at Long Island Sound by the fairly direct route of a mountain brook, the White River, and the Connecticut River. If it came down 18 inches to the north and bounced off my left shoulder, it would have to find the North Atlantic via the much longer route of the Mad River, the Winooski, Lake Champlain, the Richelieu, and the St. Lawrence.

In Larsen's second stage, the ice now farther back, the three lakes joined to form Lake Winooski, which extended up the Winooski Valley to Plainfield. This too drained south, over the Williamstown threshold. The third stage, Lake Mansfield, which Larsen further subdivides, was a major turning point, for now the Winooski and its tributaries once more drained west, although not at first along the present lower reaches of the river. So easy is it to summarize baldly the meticulous fieldwork of the geologists who not only have discovered the rudimentary facts of formation and drainage but have pinpointed the various locations of the ice and the elevations of the water. What to my eye is a set of beautiful configurations like characters in Chinese script is to them a language they have learned to translate. To Frederick Larsen, for example, the sediment dipping to the east-southeast in a particular pattern tells exactly where the retreating glacier was at certain locations in Middlesex and Waterbury.

Lake Champlain grew as the ice melted, and at first it emptied south into the Hudson River Valley because it was blocked to the north by the retreating glacier. Eventually, however, the ice retreated far enough to the north to allow the ocean waters, rising from the melting ice sheets, to flow into the lake, bringing us back to the Champlain Sea and our whale and seals. We also are brought to the arrival of the first humans—a later part of the story. But Vermont's ocean experience was short-lived. Freed of its ice, the land rebounded enough to close off the sea after two thousand to twenty-five hundred years, and the streams flowing into the basin from all sides turned saltwater into fresh. It is said, though, that briny water can still be found in deep wells in the Champlain Valley.[20]

One of the ambiguous pleasures of geology is the way it jolts our ordinary conception of things, threatening our self-importance with the brevity and, some would say, the fortuitousness, of our existence, but arousing awe with profound transformations, with great magnitudes of time, space, matter, and energy. Glaciers play their part. If the vision of two miles of ice over the Winooski Valley isn't enough, consider that almost two-thirds of all the fresh water in the world is held by the current Antarctic ice sheet. If this continental glacier alone, excluding the smaller but significant Greenland ice sheet and the many mountain glaciers around the world, "were melted at a suitable rate it could feed (1) the Mississippi River for more than 50,000 years, (2) all the rivers in the United States for about 17,000 years, (3) the Amazon River for approximately 5,000 years, or (4) all the rivers of the world for about 750 years."[21]

7

I Take to the Water

Rub-a-dub-dub, three men in a tub. —OLD NURSERY RHYME

In my case there is only one man, but he is indeed in a tub.

On my early trips north, I took note of the towns along the way, the political communities that education and long experience taught me to consider as the fundamental divisions of the land. Now I am beginning to think of myself as moving through a network of watersheds: the natural communities of Salmon Brook in my neighborhood of East Dummerston, the Connecticut River, the White River, and finally over the big divide into the Winooski watershed, each of these, even little Salmon Brook, having in turn its subsidiary systems.

From a geological journey of a billion years, to the furthest reaches of which only the most abstract powers of the mind could go, I am returning for a voyage of a few hours where all the senses are engaged. I am beginning to link what I see with what I have heard and read, but as both naturalist and canoeist I have much to learn. And it is as a canoeist that I am now put to the test.

I knew something about the importance of the canoe in the history of North America but very little about its practical mechanics. Frugality and the need for a light craft led me to a nominal 12-foot fiberglass model, which with two ash and cherry paddles cost $492.45 with tax. It weighs 58 pounds, awkward to lift but easy to carry once in place on my shoulders. Lightness is a relative matter, of course. I have read of "incredible *ultralightweight* canoes," one of which, the *Sairy Gamp,* made of cedar and 9 feet long, weighed 10½ pounds.[1] Another weighed just 9 pounds, 15 ounces. On the car now, my canoe rides on foam pads and is secured by ropes.

Interstate highways slap down the hand of man with intrusive finality, but as I drive north this August day the gravelly roadside soil sports in bloom the wild carrot, which I was taught to call Queen Anne's lace; one of the many goldenrods; the black-eyed Susan; the chicory with its pure blue; and the mullein, fit totem for a phallocentric cult, whose tall, coarse spikes now bear small, yellow blossoms.

My first stop is not the river but the hilltop home of Chuck and Judy Ratté, ever generous and hospitable, who put me up for the night. They live high above Montpelier with sweeping views of the Green Mountains, west and south. I think of how many millions have been taught to marvel at the pictorial values of such a scene, the play of angles and planes, the melting blues and greens; how a smaller band of scribblers find metaphors in ground, bedrock, lofty peaks; and how a much smaller fraternity of professional investigators such as Chuck see a dynamic landscape, a play of forces from without and within, constantly shaping and reshaping the surface of the earth in a drama of vast scope and duration.

The next morning, I head to the river for a drama that is short, low-key, and slightly comic. In Marshfield, I leave the canoe by the bridge where I ended my second hike and park my car a half-mile up the road at Les Snow and Lori Baker's farm. As I walk back to the river, my heart beats faster. This will be the first voyage in my canoe. To judge by my pulse, I might be entering a whitewater race, not this placid stream. I have to force myself to notice my surroundings, the spotted Joe-Pye-weed with its big heads of purple blossoms in a wet field by the river.

I fasten a length of polypropylene rope—I must learn to say line—to each end of the canoe in case I need to secure it somewhere or lead it through dangerous rapids. I tie the extra paddle to a seat and, in the bow, I tie a small bag of supplies, including two muffins, two apples, a packet of sunflower kernels. I forego peanut butter on this occasion. The sky is lighter; the early sprinkles are stopping. I push off.

On this first day, I have no dazzling adventures, which is just as well, for I am ill prepared for them. I bought my canoe in ignorance, but like many ignorant people I had fixed ideas about what I wanted. I wanted the shortest, lightest, affordable canoe. I never asked the fundamental question: Is this the one I should buy? Some careful measurements might have indicated a different choice. This 12-footer is actually only 11 feet long with a 40-inch beam. It has two seats and a yoke, the yoke exactly in the middle. The front of the stern seat is 32 inches from the stern, the front of the bow seat (if you reverse the canoe and paddle from the bow) is 48 inches from what is then the stern. I have been advised not to paddle from the stern seat since my weight will raise the bow of this small canoe out of the water and make it hard to manage, especially in a wind. Reversing the canoe, I must kneel but a few inches from the middle. The balance is better, but the paddling is more difficult. As I head downstream in a barely perceptible current, I am dismayed to discover that with every stroke I push the canoe sideways as well as forward. How do you go straight while paddling on one side? I have read about the J-stroke and ruddering, and I practice them. With the J-stroke, which is said to be more efficient, you pull straight back, but at the end of the stroke you turn the paddle sharply away from the canoe to correct the sideways motion initiated by the stroke's main thrust. With ruddering, you hold the paddle in place at the end of the stroke but twist it slightly to one side to offset the sideways motion.

I practice earnestly, but in paddlers' lingo the canoe won't track; it won't go straight. Since I am kneeling almost at the middle, it pivots around me, the axis, like a lazy Susan. A knot of anxiety tightens in my stomach. In desperation I change ends and paddle from the stern, letting the bow rise as it will. Even here I am forced to kneel only 18 inches or so from the middle, so the canoe tracks better but far from perfectly. I discover something else. The farther back the paddle is in the water the less it pushes the canoe to one side and the better it performs the J-stroke. An unhappy thought dawns upon me: The longer the canoe, the more easily it will track, for the paddler will be farther from the middle. I never asked and was never told. I had found a short, light canoe. And there is another problem. I can sit on the seat and paddle like an old man amusing himself on a pond or I can kneel and paddle with more force and control. But my canoe is small all over. I can't rest my butt against the seat while kneeling because there isn't room under the seat for my feet without cramping them sideways; clearance under the stern seat is 7½ inches, under the bow seat 6½. I move forward, still kneeling, and find I can keep it up for long stretches. When I tire, I sit. The canoe tracks a little better; I'm getting the hang of it.

My spirits revive, and I find that I'm hungry. I pull up on a sandbar to eat a muffin and some sunflower kernels. Freed of my preoccupation with the canoe, I see deer and raccoon tracks in the sand and, on the bank, the wild cucumber or balsam apple with its maple-like leaf and large, soft fruit in a prickly skin. The spotted touch-me-not or jewelweed grows here too, its pendent seed pods like orange jewels waiting to pop and scatter their seeds if you touch them. Resting and looking, I think that I might possibly become fond of my canoe—graceless tub though it is—the way people become fond of the instruments and machines that serve them.

When I return to the water, four half-grown ducks paddle about 50 feet ahead of me. If the gap narrows to 25 or 30 feet, they fly around the next bend. After keeping up our game of tag for hundreds of yards, we come to a wide curve where I keep to the right while they stay to the left, then paddle back upstream. The river varies from 20 to 40 feet in width, and to feel safe to reverse directions they need the space they have maintained all along. I think how differently we would perceive the world if we could see with the ducks' eyes, examine the inner process by which they have maintained the minimum distance with such consistency. No doubt a similar response governs the great blue heron I surprise as I round a bend. He rises from the river bed and flies off across the valley.

Although there is a discernible current and even a few ripples, the terrain is very gentle. The river bottom is mainly silt, clayey silt, or sand. I have much to learn about stream dynamics, but the bends show characteristic erosion on the outer side of the curve, where the current runs fastest, with deposits called point bars on the inside where it runs slowest. (My lunch-time sandbar was a point bar.) The more rapid the current, the greater its carrying capacity. Even on this small stream, the eroding banks and point bars are striking enough to show the shaping power of water, especially at flood times. Bend after bend shows freshly exposed soil where no vegetation can take hold. This is partly a natural process that would occur if no humans had ever reached the valley, but it is exacerbated by the cleared land, which lacks a riparian buffer of shrubs and trees to retard erosion. Later I am to learn much more about stream stability and instability, the phenomena studied with considerable precision by fluvial geomorphologists. Even now, however, I recognize the loss of soil, the growing ox-bows.

I pass through a stretch of water where woods and hills rise above the river banks. The bottom is stony, bedrock pops up here and there, large glacial erratics are scattered about. I often have to walk my canoe, for it drags and bumps if I stay in it.

For all the tameness of the river in this season of low water, it offers enough force to teach me something. Although I can walk through the rapids, my legs offering little resistance to water of this depth and velocity, when my canoe turns sideways against a rock, its broad expanse feels the heavy pressure of the current. I can pull it free, but a stream only slightly more vigorous would pin it there or swamp it.

I pass under a covered bridge, then leave the water at the next bridge to scout the remaining distance to the village of Plainfield, where I know there is a dam. I am one man with two modes of transportation; I have to play leapfrog with myself. I hike back to the car and drive on to the village, noting that by road I have gone only 3½ miles. It took several hours canoeing and wading, one hour walking, five minutes driving, and for a jet plane overhead, the elapsed time would have been about twenty seconds. Is there such a thing as a natural human pace? What do we miss and gain? Flying, I would not have seen the clusters of wild asters along the road, or the hedge bindweed with its white blossoms and arrow-shaped leaves. But I could have looked out on miles of rumpled clouds and down on mountains, lakes, and rivers, a sight so common in our well-traveled age that passengers doze or watch movies rather than look at what human beings until two or three generations ago could only dream of seeing.

As in many old river towns, some of Plainfield's early houses were built right on the banks so that if you stand on the bridge and look upstream, you can imagine a gondola passing and ladies on balconies, taking the afternoon sun. Only there are no gondolas, no balconies, no ladies to be seen. The fancy passes, but how will I pass? Should I paddle between the houses? It would be almost like walking through their living rooms. Then there is the small dam; leaves float slowly over the lip. Could I paddle close, then haul my canoe up the bank? The bank is high and steep, and what if the hydro plant in Marshfield starts generating at just the wrong time? How much would the release raise the water here? It would be embarrassing to come to grief on my first voyage, to plunge over the dam in a tangle of lines, arms, legs, and paddles. I head back up the road.

I hike back to the canoe. Thieves have not discovered it, not this time, and after some easy paddling I reach my take-out point and jockey the canoe onto my shoulders. One hundred pound bags of grain were nothing when I worked on my uncle's farm in high school. Fifty-eight pounds are a trifle—unless you spread them over an unwieldy 11 feet and climb a brushy, rock-strewn bank where there is no path. With much puffing and grunting I make it to the car.

Canoes were as important to some Indian tribes as automobiles to us, and to the voyageurs of the French enterprises in Canada, and later the English, they were the trucks and trains. Up to 40 feet long, they could carry 5,000 pounds of furs or other cargo plus the fourteen men who powered them.[2] Driving home, I think about the obscure place of my stubby, broad-beamed, recalcitrant craft in this history of elegance and utility, and decide that it should seek immortality instead as a humble companion of the ships of mythic undertakings. Jason had his *Argo,* Ahab his *Pequod.* I now christen my canoe *The Tub.*

8

The Deer Hunter

Such incandescent energy and alertness [said of the white-tailed deer].
We pass life back and forth—the fire that burns inside us all—creating a spectacular network of interdependence.

—RICHARD NELSON, *Heart and Blood: Living with Deer in America*

I feel an ancient, ingrained call to hunt.

—WARNER SHEDD, "Wildlife Matters," *Vermont Outdoors Magazine*

In my time travels with the geologists I looked mainly at landforms, but at the periphery of my vision there passed by a stream of living beings from the simple organisms of remote antiquity, through the awesome dinosaurs and their long supremacy, and on to the magnificent mammals, some of which survive today. That life should have arisen at all is sufficient cause for wonder, its origin and essence still eluding science, but its complex and multiform development staggers the imagination as powerfully as the convulsions of the earth. Staggers and saddens, for most of the great creatures of the past have vanished. Although I take comfort in the thought that, whatever the ultimate cause, one evolutionary road led to us (at least for now), I can't forget the extraordinary array of species that Mother Nature pushed off the path and, of course, those that humans have consigned to oblivion.

There have been a number of mass extinctions, one of the most hotly debated the destruction of the dinosaurs and their kin some 65 million years ago. Was it caused by a single asteroid or comet striking the earth, raising dust clouds that cooled the atmosphere and upset the food chain? Or was it the more gradual cooling from the dust clouds of comet showers and widespread volcanic activity? Another debated extinction was played out following the retreat of the last glacier, with some scenes in the Winooski Valley.[1] As the ice melted, it left behind a tundra ecosystem much like that of northern Canada today, with various low-growing plants, some of which can still be found on the peaks of Mount Mansfield and Camel's Hump, and among the animals the wooly mammoth, remains of which have been found in Vermont.

As the climate warmed, trees and many animals made their way north. Mastodon and elk remains have been found in the state, but it is likely that the following animals also prowled the Winooski Valley: the giant beaver (as big as a black bear), the stag moose (the size of our moose), the giant ground sloth (3 tons, 20 feet nose to tail), the American cheetah, the American lion (at 1,000 pounds much bigger than modern lions), the dire wolf (bigger than the gray wolf), the sabertooth (the size of a modern lion), and the short-faced bear (one species of which, at 1,500 pounds, is said to have been "the largest meat-eating mammal ever to have trod the Earth," although the polar bear and the coastal brown bear of Alaska occasionally reach this weight). By some 12,900 years ago, most of these animals had disappeared. According to some researchers but disputed by others, they were hunted to extinction by the human beings who also followed the ice north.

Not all animals perished, of course, and among the survivors are those that make their homes in the Winooski Valley today: the black bear, moose, deer, bobcat, coyote, fox, beaver, fisher, and others. In my lifetime, the coyote moved in from the West, the moose came back from the North, and the fisher was reintroduced. When my ancestors first moved to Vermont in the late eighteenth century, there were still wolves in the woods; and in the year of my grandmother's birth, 1881, the last known mountain lion was shot. (Whether some have since returned is a matter of debate.) All of these and more Charles Darwin called "mere pygmies" compared to the "great monsters" of the past, for the surviving species tend to be smaller.[2] Monsters, however, are not known for their beauty, and while I regret their passing I console myself with the forms and habits of the survivors, in particular with the grace and elegance of the white-tailed deer.

Of the valley's game species, it is the white-tail that most interests the hunter, writer, and conservationist Warner Shedd, who lives in East Calais in the watershed of the Kingsbury Branch, tributary to the Winooski, and who has made deer hunting a family tradition. In 1972, his son Mark, then twelve, shot his first deer, "a beautiful eight-pointer," with his father's rifle and for Christmas received the barrel and action of a .30–06 Mauser ("thirty aught six," hunters say).[3] He later bought a rough walnut stock with his own money, fitted it, and worked the finish "to a satiny gloss." Five years later an identical barrel and action appeared under the Christmas tree for his younger brother Dave after his first deer. There are New England hunting families more famous—the Berniers of Maine, the Benoits of Vermont, Larry Benoit often acclaimed as the best white-tail hunter ever—but the Shedd family brings together in the valley all the ingredients of the tradition at its most interesting: family affection and pride, a love of fine instruments, skill, respect for the game, an understanding of the complexity of wildlife management, and a land ethic based on an informed concern for habitat.

Warner has written of these things in his books, *The Kids' Wildlife Book* and *Owls Aren't Wise and Bats Aren't Blind: A Naturalist Debunks Our Favorite Fallacies About Wildlife* (an excellent introduction to some forty animals and birds), and his many articles for *New Hampshire Wildlife, Vermont Outdoors Magazine,* and its successor *Outdoors Magazine.* After twenty years as New England Regional Executive for the National Wildlife Federation, the nation's largest private conservation organization, he took early retirement to write, for "Words are my Tinker Toys, my Lincoln Logs. I would rather play with words than do almost anything else, anything indoors anyway."[4] A number of his articles relate the adventures of this little band of hunters—Mark, Dave, a hunting companion, and Warner himself, the mentor and paterfamilias—and love and pride of family and friends are as evident as the excitement of the chase. His wife Edie doesn't hunt, but she was a partner in deciding when the boys were ready for their first rifles. When Mark shot his first deer, Warner said, "I grabbed him in a bear hug and swung him around," then checked to see if the "Mighty Krag" was safely unloaded and the buck properly tagged: "I was even more proud of him for remembering those things, and I told him so."[5] It is not only the habits of deer that attract Warner's eye but the characteristics of the hunters, such as Mark's "undue optimism about approaching deer seasons," or their friend's luck of "almost unbelievable proportions," which led Mark to say, "He is not of this world."[6]

The Mighty Krag had a history of its own, and when I spoke with Warner in his comfortable log home in the hills of East Calais, I was drawn back to my early years when my father and I, then hunters, bought and talked guns. To people who have never handled firearms, it may seem odd that instruments designed to kill should participate in the elegance of all fine contrivances such as canoes, sailboats, clocks, or violins. But such is the case. A noted paleontologist has written glowingly of "Folsom-style points," spear heads from over 12,000 years ago whose "manufacture was a labour of love, for few people anywhere have ever crafted such elegant artefacts from stone."[7] For Warner and me that afternoon the fine contrivances were guns.

As with the geologists, I was soon plunged into an arcane world of technical considerations, which baffle the outsider but give flavor as well as substance to the expert. The Krag, originally known as Krag-Jorgenson, which earlier belonged to Warner's uncle, was an old army rifle from the era of the Spanish-American War. It had been sporterized (the barrel shortened) and had a broken stock. The uncle gave it to Warner, then newly married and poor, who scraped up enough money to order a rough walnut stock, which turned out to have a "gorgeous fiddle-back grain," so named because it is the grain often found on violins. After shaping and sanding, he finished it with Herter's Liege, which he now suspects was very similar to spar varnish, a petroleum distillate carrier with suspended solids. (As a boy I dreamed over the Herter's catalogue, a treasury of gun-related things.) The secret is less with the Liege itself than with the many coats, repeatedly smoothed with wet sandpaper or steel wool, followed by mineral oil and rottenstone for a satin sheen. Warner mounted a scope and blued the barrel with bluing from Herter's; bluing produces a barrel's deep blue-black, a color that pleases the eye of a hunter almost as much as a stock's rich walnut.

The caliber is .30–40, a cartridge 0.30 inch in diameter and, in the early black powder days, loaded with forty grains of powder. When the Krag became a smokeless-powder rifle, the commercial ammunition loads were slow-moving, with heavy 180- or 200-grain bullets. (In weaponry, both bullets and powder are measured in grains, one grain equaling 0.002285 ounce.) Warner bought reloading equipment and started loading his own cartridges with thirty-four to thirty-six grains of powder and lighter 150-grain spire point or spitzer bullets, the lead shape and weight making the bullet faster, with a muzzle velocity of about 2,600 feet per second, not much slower than the classic .30–06 he gave each of his sons for their first rifle. "A very powerful gun," said Warner, "absolutely devastating to deer," which is what one wants for a clean kill. Furthermore, the Krag's bolt action is "one of the smoothest actions ever invented"—smoothest but not the strongest, which limits the load. When I spoke with him, Warner had been hunting deer for thirty-seven years, and for about twenty-seven of them he used the Krag. It has now gone into honorable retirement, its hefty 10 pounds and bruising recoil having been replaced by a Ruger .243, at 7 pounds a "lovely little gun" with a muzzle velocity of nearly 3,000 feet per second and a 100-grain bullet. The faster the speed, the flatter the trajectory (the arc described by a bullet as it travels through the air), which simplifies aiming at distant targets; and the lighter the bullet, the smaller the recoil, since for every action there is an equal and opposite reaction. "High school physics becomes useful," Warner said.

Although too modest and honest to ignore the role of luck—"The deer make fools out of us over and over again"—Warner has acquired considerable skill from his decades in the woods. Asked if he could track down a deer, he said that he has done it but never a buck, or at least never a deer he could be sure was a buck. As with the practitioners of other sports, hunters sometimes disagree about technique, and of course what works in one set of circumstances may not work in another. With all hunting and fishing, the two primary requirements are knowledge of the territory and knowledge of the habits of the prey, which is why hunting since ancient times has been both metaphor and practice for warfare.

The three most common methods of deer hunting in Vermont are stand-hunting (standing or sitting still or perching in a tree stand), tracking, and still-hunting (moving quietly through the woods). Still-hunting accounted for about half of the thirty largest bucks listed in the Vermont Big Game Trophy Program in 1987 to 1999 (the biggest weighed 269 pounds dressed—without entrails, heart, and liver).[8] In the territory they know best, the Shedds prefer a quiet form of "pushing," rather close to still-hunting. Warner dislikes the term "drive," for "it brings to mind big organized efforts where people bark like dogs and beat on pans." In pushing, several hunters move slowly through the woods, almost still-hunting, while another member of the party is stationed where it is hoped a deer might pass. Over the years, a great deal of work and thought has gone into where deer are likely to go, "But the vast majority of time they don't go where you predict." At one time, still-hunting was more productive. Warner's admittedly unscientific hunch, hard to prove, is that under hunting pressure, bucks have become more nocturnal, the surviving and breeding bucks tending to be those that move more at night.

Warner's aversion to raucous drives, which bear more resemblance to herding than hunting, is just one expression of a pervasive hunting and wildlife ethic. Although he believes that human beings are hunters by nature—"like many others of my species, I feel an ancient, ingrained call to hunt"—he does not believe that this evolutionary inheritance commands indiscriminate slaughter or a callous or contemptuous attitude toward the prey.[9] A hunter who whooped "Yahoo, I got one, I got one!" disgusted him with his lack of "respect for the quarry."[10] Respect starts with the method of hunting, engaging only in what is called fair chase; continues with doing everything possible to avoid only wounding an animal; carries on at the kill with a distinction between a "sense of elation" and "gloating"; and concludes with the proper dressing of the carcass and the proper cooking of the venison, "the ultimate gesture of respect for the game." He believes in strict enforcement of the game laws and rejects as "myth" the folklore of the backwoods family kept from starving by the father's poaching: "Today's poachers are mostly in it for profit and excitement."[11]

Although indiscriminate hunting and trapping have done great damage to many species in the past before game laws were widely established and enforced, the most serious danger today isn't human predation but the loss of habitat. This may seem odd since deer are extremely adaptable, making their homes everywhere from the mountains to patches of cover in suburban and city backyards (urban rats, some people call them). The morning I wrote this sentence, my wife Eleanor called me to the kitchen window to see the hosta stripped of leaves by a deer during the night. Although they prefer a variety of wooded, brushy, and open land, providing a mix of food and cover, they thrive in many habitats, as the writer Richard Nelson tells us, in part because they "can eat and digest nearly anything that grows" even on rare occasions "animal flesh, especially fish. For example, white-tails on Michigan's North Manitou Island eat small fish called alewives that sometimes wash in large numbers onto the beaches."[12]

Despite this adaptability, "Habitat is key," as we read on a key ring handed out by the state's Fish and Wildlife Department. And fundamental to Warner Shedd's wildlife ethic is the recognition that without proper habitat, animals cannot thrive as a species and may die miserably as individuals. Starvation is crueler than a hunter's bullet.

In the warm seasons of the year, deer have no trouble finding food and shelter, but as the biologist John Buck (felicitous name) will tell you, winter and early spring are challenging times.[13] Nothing excites as much comment by wildlife experts as the question of deer wintering areas, more commonly called deer yards, and the related question of the size of the herd. The single most critical condition for the health of the herd is the availability of adequate wintering areas, stands of softwoods such as pine, fir, spruce, and cedar on slopes below 2,000 feet in elevation, facing south or southwest. These softwoods, known as the core area, hold some of the snow aloft until it evaporates, making walking much easier, and they shield deer from the winter wind. Surrounding the core should be hardwoods or mixed stands providing access to the warming sun on clear days and to the browse (twigs, small branches) that makes up much of the winter diet. Big yards with lots of deer offer more protection from predators than small yards with fewer animals.

If food is plentiful in the summer and fall, deer enter winter with the accumulated fat that is essential to their survival. Their metabolism slows, "akin to a walking hibernation," and under ideal conditions they move as little as possible.[14] Lying down, it is said, may require only two-thirds as much energy as walking around. Deer can go for weeks without appreciable amounts of food as long as they are sheltered and aren't stressed. If the deer are disturbed by dogs, say, and forced to run, they will quickly use up their stored fat. Twigs and bark have little nutritional value, and deer have been found starved to death with their stomachs full of browse, perhaps low-quality browse such as spruce. (Feeding deer won't help unless started early in the season because in the winter their digestive microorganisms can handle only browse, not grain or hay.) Oddly enough, the aggressive, sex-driven buck may be most at risk, losing 10 to 20 percent of his fat in late fall, for during the rutting season mating is more compelling than eating. But even does may lose 15 percent of their weight over the winter. A loss of 30 percent is often fatal.

As is obvious, the adequacy of habitat is determined not only by the amount of appropriate land but also by the number of deer, and human intervention is critical, for good or ill. The Indians relied on the deer for many purposes, using a wide variety of hunting methods including controlled burning both to drive and capture animals and to rejuvenate brushy forage. They also appear to have practiced "management" in the light of what we might call a conservation ethic, maintaining the vital resource (see chapter 21, however). They "regularly used fire to manipulate their environment," according to Richard Nelson, "a practice so widespread that it altered the ecological face of pre-Columbian North America."[15] According to very rough estimates, there were 2.3 million Indians in the United States and Canada (some writers say many more) who, before the days of fur trading and market hunting in collaboration with whites, took 4.6 to 6.4 million whitetails per year. This was out of an estimated deer population in 1500, just before the European influence, of 23 to 34 million. It is likely that deer numbers were fairly stable or fluctuated slowly over time.

The next 300 years radically undermined this stability as Indian trade with whites, as well as white hunting, reduced the population to an estimated 35 to 50 percent of its previous size. From 1800 to 1850, deer nationally made a healthy comeback, although not to the pre-1500 level, because Indian hunting was largely eliminated in whitetail territory, and abandoned farmland in the East and South went through a period of regeneration, producing favorable brushy and new-growth habitat. But then began the fifty-year period of massive market hunting, mainly by whites with modern rifles, "the period of greatest hunting pressure on wildlife *ever.*"[16] Big cities in the East and Midwest were ready markets, increasingly accessible because of railroads and even early refrigerated cars. In 1860, 6,000 deer were killed in Minnesota by a father and son team alone; in 1880, 70,000 were killed in Michigan.

By the end of the century, the joy-ride was largely over, the herd in the United States now reduced to about 300,000. It was at this time that game laws began to grow teeth. Earlier regulations, some dating back to colonial days, had been generally ineffective, but a rising environmentalism and, in particular, sport hunters' concerns over the fate of the deer led to support for new laws, more strictly enforced. "So it was mainly hunters and fishermen," Nelson writes "who formulated a new conservation ethic, who became the first and most prominent voices for responsible use of wildlife, and who advocated preservation of natural habitats."[17] Warner Shedd remarked that the term "sport hunting" is perhaps unfortunate because to some people there is nothing sporting about shooting a deer with a high-powered rifle, but the term came into general use at the end of the nineteenth century to distinguish the person who hunted for personal pleasure and use in the spirit of fair chase from the market hunter who might mount a gun on the bow of a boat and kill fifty ducks at once. And Lowell Halls writes, that "the history of the white-tailed deer helped to stimulate the concern that was [the] origin of modern wildlife management, initially as a political process of trial and error and eventually a scientific discipline."[18]

In the Winooski Valley and Vermont generally, the nineteenth-century story was different. As Vermont's *White-tailed Deer Management Plan (1997–2006)* explains, instead of rebounding from 1800 to 1850, as it did in many places, the deer population in Vermont was lowered nearly to zero because of hunting and the loss of habitat; in less than a century, farming and logging had reduced the state's forest cover from 85 to 30 percent.[19] (In his *Natural History of Vermont,* published in 1853, Zadock Thompson said that deer, once "one of the most common and valuable quadrupeds found in our forests," had shrunk in number "till they have become exceedingly scarce, except in a few of the most unsettled and woody sections."[20]) The management plan was drafted in 1997 by the Deer Management Team headed by John Buck and is the source of most of the information that follows.

Given half a chance, deer are remarkably resilient. The half-chance for Vermont's herd was a combination of events—the reintroduction of deer from New York, a ban on all hunting from 1865 to 1897, the conversion of abandoned farmland to new-growth woodland—that led to an 1897 bucks-only hunting season with a harvest of 103. By 1940, thanks to more abandoned farms and limited seasons, the take rose to more than 4,000. Healthy deer reproduce rapidly. While single births are usual for one- and two-year olds, twins are the rule thereafter for as long as ten years, and triplets are not rare. Wildlife writers often cite the experiments conducted on the George Reserve in southern Michigan, a fenced, 1,146-acre range where deer have been studied since 1928.[21] In one experiment, ten deer increased to 212 in five years, 1975 to 1980. This reproductive energy fueled the growth of the Vermont herd until in 1946 the state's biologists were concerned enough about habitat destruction to recommend doe seasons. Even now, though, some hunters think that killing does keeps the population below what it should be, and I can remember the general antipathy when I was a boy and young man. Public opinion won out in the legislature, beginning what the *Management Plan* calls "an infamous period of deer management in Vermont." The population grew but became less healthy, and habitat continued to deteriorate. Modern management began in 1963, but the timid antlerless seasons to 1970 took less than 5 percent of the estimated herd of 250,000. (Remember that the entire whitetail population of the United States in 1900 was estimated at 300,000.)

A bad situation was about to turn worse. Vermont winters are erratic and unpredictable. In 1969 and 1970, two severe winters in a row killed more than half the deer, cutting the herd to 120,000, and the habitat was so damaged that it could not support the normal reproductive rates of even this much-smaller number. Although largely beyond my scope, the political dimension of deer management from the 1950s to the present is a story in itself, in Warner Shedd's words a hot potato for the legislature. It was hard in the early years for many hunters to see the need for long-range balancing of habitat and numbers. The quiet words of the *Management Plan*—a goal of "a proper balance between adequate numbers for human enjoyment, tolerable nuisance situations, and ecological integrity"—give little hint of the noisy battles that actually took place. And it is important to note that there is, ultimately, a democratic constraint on what the biologists might want to do for the good of the herd, for as the Department's mission states, the goal is the "conservation of fish, wildlife and plants and their habitats *for the people of Vermont*" (my emphasis). If wildlife management is not to succumb to the passions of the moment, education and persuasion must bridge the gap between the judgment of the biologists and the immediate desires of the people. Experts are not always right, but Warner describes Vermont's biologists as "thorough-going professionals": "If I'm ill, I go to a doctor and so would most people, but when it comes to wildlife, everyone is an expert. Except me."

Despite setbacks, the Department pushed on with its work and in 1979, as the *Plan* says, "began an ambitious deer population recovery effort" that first reduced the herd even below the winter-kill level, then stabilized it for the sake of the habitat, and finally increased it slowly with the goal of annual harvests of 15,000 to 20,000. The program was working, but hunters had to endure radically reduced yields in the 1980s, and the legislature in 1990 banned November doe seasons. At this point, the Department began a more ambitious effort to solicit and educate public opinion. A management plan for 1990 to 1995 was drawn up that divided the state into twenty-four areas now known as Wildlife Management Units, established buck harvest goals for each unit, allowed the taking of antlerless deer during muzzle-loader and archery seasons, and tried to recruit more hunters.

The work of the Department has been greatly assisted by heavy reforestation, mainly due to the continuing abandonment of farms. Eighty-five percent of the state is now wooded, although the figure is somewhat misleading in that many wooded stretches are dotted with houses. I asked Warner what he considered the greatest threat to the deer herd today. It is, he said, "a house here, a house there," small-scale development that pushes deeper and deeper into the woods with more winter activity, including the prowling of dogs, to stress the deer. Since many towns do not have wildlife protection as part of their town plan and zoning ordinance, deer yards are open to invasion except for developments large enough to call into play Act 250, Vermont's main conservation law, or the rules to protect wetlands.

Later the Department established a "youth hunting day" (now a weekend) allowing hunting before the regular November season, an "Outdoors Woman" program encouraging women's participation in outdoor sports, and a "Working for Wildlife" program sponsoring cooperation with sportsmen's clubs in various habitat and conservation projects. Details differ somewhat, but the basic principles of the 1990 to 1995 plan and these subsequent efforts are incorporated in the 1997 to 2006 plan with the added provision that so-called quality deer management could be attempted in any management unit if there was sufficient hunter support. Methods vary, but quality management reduces the take of young bucks to allow more of them to grow bigger and to sport larger racks. The trade-off, of course, is that hunters have a smaller chance of getting a deer. There are no quality management units at present. (The Benoits no longer hunt in Vermont because of the paucity of big, mature bucks.[22]) Currently there are three seasons—any legal weapon, muzzle loader, bow—with a limit of three deer per hunter in a calendar year. (These regulations are subject to continuing review and modification.) When the object is to take only bucks, the actual criterion is antlers rather than sex because although normally only bucks have antlers, occasionally a doe will have them or a buck will be bald.

Each year the Department publishes a report on the most recent deer seasons including weather information and detailed figures on the number of licenses sold, the number of deer taken by season and location, the percentage of successful hunters, and biological data. You can't manage without information. A comparison of 2000 and 2001 provides a snapshot of management issues.[23] The 2000 season, Vermont's 104th in a row, followed a number of mild winters, low to moderate antlerless harvests, and abundant fall food. The fall population was estimated at 150,000 to 160,000, and the total take was 20,498; 17 percent of the hunters took at least one deer. Heavy snows in March and April of 2001, however, killed an estimated 20 percent of the herd, fewer antlerless permits were issued, and the total take was 15,065; 13 percent of the hunters took at least one deer. The harvest in those towns with roughly half or more of their area in the Winooski Valley dropped about 35 percent, the harsh, mountainous terrain perhaps accounting for the greater than average decline.

I spent a morning at a check station with John Buck as he and his colleagues gathered information from deer taken the opening day of rifle season. They weighed each deer, measured the antler beam, counted the points, checked for ticks (found none), and looked at the teeth to determine the animal's age.

The beam is measured by taking the average of the two diameters (at right angles to one another) an inch above the burr or base of the antler. As John said, the beam gives "an indication of the nutritional plane and genetics and overall health of the individual animal. We put these figures into a pool of all animals of his age." Comparing the results of various years, John can see if a trend is developing in health and habitat that might require some management adjustments.

John counted the official points; to be official a point must have grown at least one inch off the main beam. A so-called hunter's point, as the joke has it, is simply one long enough to hang a ring on. Then John checked the teeth to determine age. He is 100 percent accurate with yearlings because of baby teeth or their replacements. With older deer he is 90 percent accurate, judging the degree of wear. Yearlings are the most common. In any year, many more does than bucks will be ten years or older, partly because not as many does are shot but also because does are less rambunctious and reckless: "They're not running all over the hillsides as bucks do in rut, running into cars, into fence lines, into each other, whatever." Even as a fawn, the buck is more likely to stray from the mother and be picked off by a coyote.

The smallest deer this morning weighed 106 pounds. The biggest weighed 176 with a 23 × 32 millimeter beam and a grand, eight-point, spreading rack. He was four and one-half years old.

I grew up in a time when nearly everyone accepted hunting as part of Vermont's way of life. For whatever reason—newcomers with different attitudes, a shift in opinion among native Vermonters—the number of hunters has declined over the years. The Department is concerned, for license fees raise money for wildlife programs, hunters directly benefit the economy of a number of towns, and hunting is a critical management tool. While it isn't true in every corner of the United States, in Vermont there is no feasible and humane alternative for controlling the size of the herd and protecting the forest. When I think about the harsh consequences of too many deer, I feel guilty that for years I haven't hunted. Hunting is kind.

In fact, it appears that not only numbers but also the physical characteristics of the species have been determined to a significant degree by human impact. You can imagine a pristine state of affairs with no human intruders in which deer, animal predators, and habitat would achieve a rough balance, fluctuating only according to changes of weather and climate. But the deer as we know it is probably not the product of a purely natural, that is, non-human world. The Indians or their predecessors followed animals north as the glacier retreated, and they depended heavily on whitetails well into modern times for food, clothing, tools, ornament, and ceremonial purposes, as many early Europeans in North America observed. Some writers believe that human hunters did even more than natural predators—bobcats, mountain lions, wolves, and coyotes—to produce the "evolutionary sculpting" of the deer: the remarkable adaptive characteristics of keen senses, nervous attentiveness, coat color, speed, and agility.[24]

To look at the details and language of a profession is to see the world through different eyes. The only hope for the survival of a healthy deer herd or indeed for many wildlife species, plant and animal, is through management broadly understood. With an increasing human population and the physical and economic means to plant ourselves in ever more remote areas, management is critical. But wildlife management is based on two disciplines: the biological sciences especially as they contribute to ecology (organisms in their environment) and the theory of management as control. To me, the inquiring outsider, ecology is a privileged glimpse into other lives as they are actually lived. Management as control assumes an instrumental relationship to the thing managed. Its language, as applied to wildlife, owes much to scientific agriculture. Deer are a resource, like corn, to be harvested. Grateful as I am for the necessary work of these dedicated professionals, there is something in me—and I suspect in them—that isn't satisfied by the notion of a completely managed wilderness, although for much of the world I see no alternative. In fact, a managed wilderness is no longer truly wild; it is to a greater or less extent a human construct. The goal of management is, in part, to maintain a "wild" environment suitable for the species being managed, but in the way we think about them the species themselves cease to be wild or independent beings.

It would take me too far afield and into some murky realms of thought and feeling to try to explain in detail why "wildness" matters to me. Part of the explanation may be nothing more than the experiences of boyhood when I went into the woods, gun in hand, thinking not of the fact that fifty years ago it was a farmer's pasture and that it was ultimately subject to human control, but imagining, in contrast to my nearby village, a place utterly different, a place with its own rules and creatures, only partly understood and therefore fascinating and even a little dangerous. The managers I came to know and admire reminded me of the naturalists of another era. Although they spoke the language of the professional managers that they are, they saw animals as living beings in a complex environment, not simply as quantities of raw material. I can't speak for them, but in listening to them talk about the creatures they were managing, I imagined that I heard overtones of something like love, something that transcended the utilitarian analysis of resource and yield.

9

In Search of the Wild Onion

Many families have lived for weeks on what ye people call leeks, a sort of wild onion, very offensive to me.

—THE REV. NATHAN PERKINS, *A Narrative of a Tour through the State of Vermont*

I dug out three with my fingers—moist, dark soil—deeply, tenaciously rooted.

—MY TRAVEL LOG

Now christened, her maiden run behind her, *The Tub* is ready for her second voyage, the one technical improvement a kayak paddle to help a stubby craft and her novice master cleave a straighter track. I put in at the Plainfield Recreation Area and have 50 feet of relative calm to practice in before hitting a long stretch of intermittent rapids. Although not very threatening, twice they turn *The Tub* end for end. I can't remember what to do, but when you are bouncing backwards down a river, you improvise, you learn fast. Now and then I hang up on rocks. My first impulse is to look around to see if anyone is watching. Relieved to be alone, I step into the cold water, feel the constriction in my chest, and think about having a heart attack in the middle of a river. Gradually my feet adjust to the cold, and I walk *The Tub* away from the rocks.

I recall my meeting with the naturalist Charles Cogbill of Plainfield, an ecologist with a special interest in landscape history and historical ecology. "He knows everything," I was told, and indeed he referred me to several useful books and told me where to search for the wild onion for which the river is named. Looking something like a lily, it has wider leaves than cultivated onions, and I keep an eye out for it as I make my way downstream. I may be too late; Charles says that in June the leaves start turning brown.

The gradient flattens out at Route 2 as I cross into East Montpelier. I enter former farmland, and the river here for an amateur canoeist is a perfect stream with a gentle current and inconsequential rapids. It's the most pleasant reach above Montpelier, but I am learning to look with new eyes, and what I see tells a tale of human folly and nature's power. Stream dynamics is the name given to the way streams act on the land, and there is much here to suggest that dynamism strikes just the right note. Streams change their courses over time with or without human presence, but the process is much accelerated when water velocity soars because of rapid run-off from cleared land and when stream banks have been stripped of protective vegetation. The river banks here have seen significant slumping, falling into the river as the current undercuts them, and some sections qualify as mass failures, the geologists' term for a more extensive collapse. Here I find great fallen chunks, some resting in the water, some with green grass and trees. From 8 to 25 feet high, the torn banks go on for 200 to 300 feet, and there are big sand and gravel bars on the inside of bends. Later I will see superimposed aerial photographs of this stretch of river from 1942, 1979, and 1988.[1] In the forested section upstream of the Route 2 crossing, the river kept to the same channel; downstream of the highway, without an adequate riparian buffer, it was another river altogether, cutting new channels, moving laterally in some places as much as 400 feet. The tillable land lost to a buffer of trees and shrubs would have been much less than that lost over the decades to a freewheeling river.

A rope hangs from a tree over a deep pool, waiting for kids when the weather warms. Red-winged blackbirds take their short, erratic flights, swallows dart and swoop in pursuit of insects, and sandpipers hop along the stream edge, teetering on their slender legs. Higher up the food chain, human predators are at work too. A silver-haired aristocratic couple sporting tackle and regalia—rods, nets, hats, vests, waders—out of Orvis or L. L. Bean, cast their flies in a poor stretch of water as if their gear outstripped their knowledge. At the mouth of the Kingsbury Branch farther on, their rural counterpart, a young man with spinning rod and worms, says that he catches browns and rainbows, but right now a small fish is stealing his bait. He is wearing blue jeans, no hat or shirt, and his red hair and white skin stand out against the grassy bank. He is my reincarnation from fifty years ago. I apologize for paddling through his fishing hole, but he isn't bothered: "Everyone has a right to his sport."

As I learned earlier, this area of Plainfield and East Montpelier has two sites of interest to Chuck Ratté, launch pads from which he propelled me into the past. Geologists owe much to human construction and destruction, road cuts and the like, and here the work is two man-made pits. The first is an abandoned clay pit just west of Plainfield village off Route 2. (The clayey till may have been used to fill in the low spots of a farmer's field.) "The glacier stagnated," Chuck said. "It simply sat and melted," dropping so-called ablation till, which is distinguished from the lodgment till laid down under the ice as it advanced. Composed of all sizes, clay to boulders, the debris formed a big sheet or bank known as a ground moraine that lies under the village.There was no opportunity for streams to sort and wash the jumbled particles. This condition isn't the same as a moment of equilibrium in which a glacier doesn't advance because the front is melting at the same rate as it is moving forward, a condition that creates a linear bank of till known as an end moraine. Chuck pointed out the yellow trefoil, crown vetch, purple vetch, and red clover probably planted to control erosion after the pit was closed.

Chuck noted that the area west of Plainfield village where the river crosses the road and enters East Montpelier is an "obvious basin"—obvious once he pointed it out, but I might have paddled through it unaware. This was part of Lake Vermont, formed when what we think of as Lake Champlain, blocked to the north by the retreating glacier and gathering water from the melting ice and many streams, backed up the Winooski Valley and elsewhere in a high-level lake. A stream or streams flowing into Lake Vermont in this area sorted the till and washed out enough fine material to form a delta in the lake.

For evidence of this, Chuck took me to the second site, an abandoned gravel pit on a side road just north of Route 2. We climbed high up the steep side of the pit, where disturbance was minimal, and with a jackknife Chuck delicately cut a fresh vertical surface. There stood revealed, like a photo transported by time travel, the beds of the glacial lake bottom as they were laid down some 12,000 to 14,000 years ago, only slumping a little over time. We were looking at varves, pairs of sediment layers of alternating degrees of fineness. In the winter, with stream action and lake water movement at a minimum, particles would advance little if at all and the finest clays would have a chance to settle out. In the summer, with more activity, all particles would be carried farther, clays would be kept in suspension, and coarser sediment would be deposited. Chuck told me to rub a pinch of each layer between thumb and finger. The tannish brown silt was slightly gritty, the gray clay quite smooth.

Still in the gravel pit, we found a rough, chunky piece about 15 inches long, a jumble of different rock particles that Chuck identified as quartz, schist, serpentinite (from northern Vermont), hornfels, and others. It was a sample of modern cementation of the glacial gravel deposit, "modern" meaning any time from the late glacial period to 5,000 or 6,000 years ago, or even more recent. The cement is calcium carbonate. Where did it come from? Underlying this whole area is the Waits River bedrock formation, consisting of impure marble. As water works its way through cracks in the bedrock marble, it gradually eats away at it, dissolving out the calcium carbonate, holding it in suspension, and transporting it to gravel deposits at a lower elevation. The calcium carbonate re-precipitates in the gravel, cementing layers that can be broken into coarse chunks like the one I held in my hand.

I am paddling, as always, over ancient history, but now I think of contemporary inconvenience. As I discovered the first time out, I can't lean against the thwart; it's too low and my feet would have to twist to fit under it, dangerously locked in. I find, though, that I can kneel erect for long periods, and when I feel lazy or my knees start to ache, I sit on the seat. Sitting makes for awkward paddling and is very unprofessional, but nobody's looking. I paddle and drift. Time doesn't matter, a rare release for me, for I am as time-driven as any modern man. In a perfectly calm stretch, *The Tub* goes with perverse accuracy to a submerged log and sticks there. Grateful not to be observed, I push and rock until I'm free. I go under a covered bridge and a bit later under Route 2, having now navigated the big loop north of the road. Farther still, I stop to rest and to eat my bagel and sunflower seeds—no peanut butter this time. Across from where I am sitting, an eroded bank some 10 feet high and 200 feet long is punctuated by about fifty swallow holes. Swallows are twittering and circling around the bank with momentary stops at their nests. Suddenly the entire cloud rises up and disappears downstream. By what signal or instinct do they move in unison? I assume there are young ones in the nests. I am tempted to look, but I refrain from disturbing them.

I come out of the river at the next Route 2 bridge near Codding Road just before the Green Valley Campground. I hide the canoe under the bridge and walk back to the car, stopping for an ice cream cone in East Montpelier—a double dip of maple walnut. I drive back to the bridge, load the canoe, then park. Where I can't canoe I hike, almost always, and the next stretch of the river looks questionable with three dams and a rocky stream bed. I have not forgotten that I am on a quest for the wild onion. I have about decided that I am too late, it has gone by, when a half mile or so below the bridge I come upon a small patch of likely looking plants, then a bigger one. Once I start digging with my fingers, my nose confirms the find; there is no mistaking them. They root deep, but I excavate three, put two in my hip pocket, and eat one on the spot. It's an onion all right, sharp and tasty.

IO

From Influent to Effluent

A Clean Story and Some Dirty Ones

Escherichia coli make up one-fifth to one-third of the weight of an average feces.

—ALICE OUTWATER, *Water: A Natural History*

Bacteria feed the world.

Until 1963 city sewers ran down all the major streets and emptied directly into the Winooski River and North Branch. —HANS ESTRIN AND OTHERS, *Montpelier Rivers Report*

This pump is twenty years old and runs like a top. We hang on to our equipment.

The Winooski is a Class B river.

—RONALD MERCIER, Chief Operator, Montpelier's Water Pollution Control Plant

Since Vermont law, reflecting the federal Clean Water Act, classifies all state waters as Class B or higher, the Winooski's designation may seem a dubious distinction. Furthermore, not all of it is in fact Class B, for the *Vermont Water Quality Standards* describes what the state's waters should be, not what they are; a few years ago one-fifth of the state's river mileage fell below federal standards.[1] Add to the picture Vermont's grade of D in a water quality report by the National Wildlife Federation not long ago, and the scene looks dark indeed. But I remember when the Winooski was a running sewer. Even to aspire to Class B is an achievement, and in fact the river is much cleaner than when I was a boy.

First the *Standards*. The document divides the state's waters into seventeen basins and for each one declares that all waters are classified as B except for those above 2,500 feet in altitude and certain other named ones, which are called A. The natural purity of Class A waters is to be protected from human degradation. In addition there are mixing zones, short sections of a stream below a treatment plant where effluent is absorbed and diluted. Plainly put, Class B must be clean enough for swimming, fishing, and other recreation, which is what the federal Clean Water Act some thirty years ago set as a minimum standard for all United States waters. (Eventually all Class B waters must redesignated and divided into three Water Management types according to use, a refinement I can skip since it won't affect the Winooski River basin for several years.)

It may be obvious that despite significant progress the goal has not been met in Vermont or nationally. Federal money has gone mainly to combat major sources of pollution such as municipal sewage. Furthermore, budgetary and political obstacles have hampered the state's efforts to clamp down on small-scale polluters, especially the nonpoint sources (explained later), whose combined impact is significant, although here again progress has been made.[2]

The *Standards* also distinguish between cold water and warm water fish habitats. In the Winooski River basin, thirteen ponds are classified as warm, including Coits at the headwaters of the Winooski, as are nearly all wetlands, and the lower Winooski from the dam at Essex down to Lake Champlain (but only from June 1 to September 30). Everything else is called cold water. Curiously enough, the designations are not entirely a matter of temperature but have to do with the habitat type more broadly considered and the species of fish present.

From open sewer to mainly Class B, then, is one of the major clean stories of the waters of the state, and for the Winooski one of the main characters is Ronald Mercier, Supervisor and Chief Operator of the City of Montpelier's Water Pollution Control Plant, one of Vermont's some 150 wastewater treatment plants, municipal and industrial. It was through his eyes that I saw why, if we are not to foul our rivers, we need trained personnel and multi-million dollar facilities to dispose of what we swoosh away with hardly a thought. Foul them we did, for generations, and the consequences are worth revisiting before I retrace my tour of the Montpelier plant.

Before we work ourselves into a froth of indignation at the offenses of our ancestors, however, it is well to remember that few if any of them woke up one day saying, "I understand what happens when I pollute the river, I know it's bad, but I am going to do it anyway for my own selfish reasons." Few human beings have the self-knowledge, honesty, and courage to live that way. Most of us have to believe that the bad we do is incidental to the greater good, which is of course why we are capable of such enormous destruction as well as such admirable achievements. The ways of our fellow creatures in the forest are intriguing and wonderful, but for the ultimate in splendid complexity, consider *Homo sapiens*. Consider the man by whose sewer pipe I sat as a boy in Essex Junction fishing for bass and noting now and then with little interest when he flushed his toilet. Consider the whole town whose sewers and storm drains like those of Montpelier emptied into the river. And so it was up and down the valley.

There were indeed consequences, but from the perspective of public health, turning the rivers of America into public sewers was at one time a great advance, at least for cities such as Boston, New York, and Philadelphia where waste disposal became increasingly inadequate as populations grew in the late eighteenth and nineteenth centuries. Human and animal waste, garbage, and other noxious material gathered in cesspools, marshes, ponds, and other low-lying areas before municipal water supplies and sewage systems were built to flush them into convenient rivers. The accumulated filth contributed to outbreaks of yellow fever, typhoid, and other diseases, some of them severe enough to bring normal city life to a halt. We read that "yellow fever struck Philadelphia with a vengeance in 1793. Almost four thousand people died, and twenty-three thousand fled, effectively closing down the nation's capital and leading port for months."[3]

Given what was known of public health and what was not known of the fragility of ecosystems, it was to be expected that as the Winooski Valley's communities grew, they would use the river as a common sewer; houses and other buildings had been dumping waste into the river from the very beginning. Equally important, I suspect, was the fact that most people had other things on their minds. Before the rise of a widespread environmental consciousness, very few people thought of their community as an integral part of a natural system all parts of which deserved consideration. What most people thought of was how to take care of their family, their school, their business, and the general well-being of their town. For 150 years or more, that well-being did not extend to the river.

It may seem obvious what raw sewage does to a river. It makes it stink, it makes it undrinkable, it makes it unswimmable (although what the health authorities consider unswimmable may by judged differently by sweaty boys on a summer afternoon). It may even make it look bad. I remember that years ago Danny Sicely could tell how high the Winooski had risen by where the toilet paper hung in the stream-side branches. But some of the effects are not so obvious. Aquatic life is a marvel of complex interactions among living beings and the physical and chemical characteristics of the stream. Disturb the stream at the lowest biological levels and you may disrupt the whole dance of life.

Where does the dance begin? Although, in a sense, cycles or webs are by definition without beginning or end, we may say that the food web begins with the sun, for the sun is the source of the energy that green plants need to transform inorganic substances into living material. The process, as we learned in high school biology, is called photosynthesis, "a series of chemical reactions that combines carbon dioxide with the hydrogen in water molecules, to produce simple sugars and release oxygen."[4] On my last visit to Coits Pond, in July, great stretches were covered with various plants, including a beautiful water lily. All of them were happily photosynthesizing, using the sugar they produced from sunlight to grow and giving off oxygen, but in the food web they are dwarfed in importance by the lowly algae that are found growing nearly everywhere in water, salt and fresh; of the green algae alone there are said to be some 7,000 species.[5] Many kinds are microscopic, but they often are visible in green and blue-green masses, and they combine with bacteria and tiny animals to form "the slick brownish film that coats nearly every object in the stream. . . . This thin film of life is the very basis of the stream community."

Algae are vital to the stream in two ways. First, they are "the river's most important primary producers" of food, eaten by so-called first-order consumers, herbivores such as snails or mayfly nymphs, which in turn are eaten by second-order consumers, carnivorous small fish, which then provide dinner for the third order of larger carnivorous fish, and on to fourth-order consumers such as herons, mink, otters, and humans.[6] The higher orders eventually die and are either eaten by scavengers or decomposed by bacteria, re-entering the food web in this inelegant but essential fashion. Obviously, the web has many routes; a very small fish might be eaten by an otter or a human being.

The second way in which algae are vital to the stream is by producing oxygen, which all living things, except anaerobic bacteria, need to live. Streams and other bodies of water contain varying amounts of dissolved oxygen depending on conditions such as purity and temperature. (Cold water holds more gas than warm water; note the foamy head on warm beer.) Tumbling, cascading streams incorporate much oxygen from the air, but even here algae are important. As a mere second-order consumer of biological knowledge, I am astonished by the variety of ways in which aquatic animals capture oxygen, but I detect a note of wonder in what even the professionals have to say. Beavers and otters breathe as we do, with internal lungs, although they can go much longer without new air. Fish take in water through their mouths and pass it over their gills where oxygen is absorbed and carbon dioxide is given off. A salamander's larva has external gills as do the larvae of many insects; a certain beetle larva cuts a hole in a water lily stalk to reach encapsulated air; other beetles go underwater carrying a bubble of air, which they replace from time to time or, if they stay under long enough, which is replenished by dissolved oxygen in the water.

Enter sewage (or any organic material such as leaves from overhanging trees). The organic material we cannot absorb is not lost to nature. When it makes its way untreated to the Winooski and its tributaries, it becomes food for bacteria that decompose it. Up to a point, bacteria can handle sewage without greatly harming aquatic life—there are fishermen who believe that some sewage treatment plants actually hurt fishing by reducing the food available to the food web—although certain pathogens in the waste itself might return to a human's system and make him sick. Beyond this point, however, things go seriously wrong.[7] Bacteria are unimaginably prolific, said to be the most numerous living beings on earth considered as individuals; there may be up to 100 million of them in a single gram of good soil.[8] Most of them are beneficial, even necessary, for a healthy environment and healthy people, but in disposing of sewage and other organic material such as masses of dead algal blooms—a good deed—they may overwhelm their corner of the aquatic world. (Algal blooms, masses of algae, are more a problem in Lake Champlain than in the river.) As the amount of organic matter increases, so do the numbers of bacteria. Since like nearly all living things, they need oxygen, in their abundance they may use up much of the dissolved oxygen in the water, depriving insects and fish of what they need to survive. They also may reduce the oxygen supply at its source, clouding the water and coating the bottom of the stream so that less sunlight is available for the oxygen-producing algae. Insect, fish, and mammal populations change with the changing oxygen level. And so it was in my hometown stretch of the Winooski where trout surely flourished before the time of heavy settlement but where, as a boy, I caught only perch and bass, more tolerant than trout of warm water and low oxygen.[9]

Ron Mercier came to the Montpelier facility in 1981, a plant that had been operating since 1964. It was a so-called primary treatment plant that left about 50 percent of the pollutants in the waste stream. By 1972, 50 percent was too much, for the Federal Water Pollution Control Act Amendment of that year set a cap of 15 percent, and the City of Montpelier began the long process of converting to secondary treatment, adding biological and chemical actions.[10] The expanded plant went into service in 1981.

The plant mimics the human digestive system but in reverse. Whereas we take in more or less pure food and excrete heavily polluted waste, the plant takes in heavily polluted "food" and excretes more or less pure "waste." Various human stages from chewing up the big particles, to chemical and biological decomposition, the production of gas, the separation of liquid and solids, and the eventual compaction and discharge have their corresponding stages in the plant. What follows is a summary of the plant's digestive system. As I followed Ron through the plant, I heard echoes of Hap Hayward's curiosity about how things work, his ease in the presence of ingenious complexity. Everything was painted and orderly, cleaner than some restaurant kitchens I have seen. "The guys take a lot of pride," Ron said. He showed me a pump that one of his men took apart, cleaned, and reassembled: "This pump is twenty years old and runs like a top. We hang on to our equipment."

Imagine a waste stream entering the plant from the sewer pipes and storm drains of the city. First, bar screens filter out the sticks, rags, and other big pieces of trash. Next, the stream enters a grit chamber where heavy particles such as sand settle out. From there it goes to primary settling or sedimentation tanks in which the heavier organic material sinks to the bottom as sludge. The sludge moves to closed primary tanks or digesters without oxygen where heat and anaerobic bacteria decompose some of the organic matter, kill many pathogens, and, as a by-product, produce methane gas, which helps to heat the plant. Bicarbonate of soda is added to maintain the right balance of acidity and alkalinity, and the temperature is kept at 95 degrees. "It's like treating an upset stomach," Ron said. "That's what it is, a big stomach digesting food." (As in a sense, so is the Winooski River, but the plant digests its "food" much more quickly.) After fifteen or sixteen days, the sludge has given up its methane, and it goes to Harold Blake, the press man, who adds polymer to make it clump and removes about 20 percent of the water through a process of thickening and belt-press squeezing. The sludge is then hauled to the landfill.

This sludge removed, the liquid waste stream continues on its journey to aeration tanks that force air into the liquid, adding the extra oxygen that bacteria and other microorganisms need to reproduce more rapidly, decomposing more of the organic solids. Polyaluminum chloride is added, which precipitates much of the phosphorus as the stream moves on to the secondary settling or sedimentation tanks. Here more sludge, with the phosphorus, settles out as activated sludge, activated because of the high bacteria content. Remember, these are good bacteria, doing an important job.

This activated sludge now divides into two parts. The waste part joins the primary sludge stream and goes to the anaerobic digesters and on to Harold Blake for final processing. The return part goes back to the aeration tanks to pep up the decomposition going on there and eventually will go to the secondary sedimentation tanks in one big circle. It requires some careful lab work, Ron said, to figure out how much activated sludge to bring back "to chew up all the new pollutants coming in. The bacteria in the activated sludge have been sitting here sleeping, and now they're hungry."

The liquid waste stream has just two more stops before it enters the Winooski with 97 percent of its pollutants removed, "a lot cleaner," said lab technician Bob Fisher, "than what many people have to drink in third-world countries." Chlorine is added as a final purifier, and then sulfur dioxide to remove the chlorine, which would otherwise build up in the river, killing all aquatic life on the bottom. In 2006, the plant will replace chlorine with ultraviolet disinfectant. Chlorine gas, said Ron, is "real dangerous; it will kill you quick."

Ron spoke of the biological controls in the plant not just as technical adjustments but as the way to maintain the proper "environment" for treatment to happen. And he wanted me to think of environmental implications beyond the immediate scope of his job. Phosphorus control, for example, is important because phosphorus is a major nutrient and therefore a major pollutant, for in excess it causes our good friends the algae to multiply until they choke out other living things: "We don't have all the green algae blooming as in Lake Champlain."

A believer in "beneficial reuse," Ron would like to see a happier fate for the sludge and excess methane. The sludge goes to the local landfill because there are not enough farms now to accommodate it as fertilizer, and a lot of time and expense is required to monitor its application and the possible effect on groundwater. Winter storage would be difficult because the plant produces three and one-half to four tons per day. And a big composting firm in New York has more material than it can sell.

The excess methane not used by the plant for heating must be burned and wasted. It's too "dirty" and "wet" to use as vehicle fuel, but with a steam boiler and generator it could be burned twenty-four hours per day to put electricity "into the grid," that is, to sell to the network of power suppliers. Ron says, "I've studied it, I did the math, it's feasible and cost-effective," and he hopes that some day the city will invest in the equipment. He processes, for a fee, septage from septage haulers and leachate from landfills. His goal is to run the plant more like a business.

Montpelier is a small city, the smallest state capital in the nation with about 8,400 people; 150 households have private septic systems.[11] When you add the state employees who come in from outlying towns, and parts of the town of Berlin and the Central Vermont Hospital, also served, the system probably handles 18,000 to 20,000 people. The state classifies the plant as Grade IV with a permitted flow of 3.97 million gallons per day and an annual average flow of 1.6253 million per day. All components, however, have a so-called peaking factor, that is, they can handle larger flows during rainy seasons or in sudden storms with very little if any decrease in effluent quality. In a combined sewer and storm drain system, an unpredictable element is the amount of water dumped by a heavy rain. Ron has seen a very heavy rain push the influent up to 11 million gallons in a day. Too much water and you get a wash-out, no matter what the operators do: "Some of the solids make it all the way through the system and you get all of that raw bacteria going through. You're putting tons in the river bottom—one flash flood." Old and new sewer lines in the city were connected to allow overflow, with some sewage, to go to directly to the river in heavy rains. Now, in a multi-year $10 million project, the storm and sewer drains are being separated and upgraded. All storm water will go directly to the river, a solution with problems of its own, but better than letting raw sewage escape. Ron estimates that with complete separation, the peak flow will drop to not more than 4 to 5 million gallons per day.

The six men at the Montpelier plant all hold state certificates appropriate for the level of their jobs. The work requires diligence and honesty as well as technical know-how. In other towns, Ron said, "They've taken a few guys to jail and fined them for not doing their work, doctoring records. There was a guy who was a good operator, but he got lax, and they walked in, caught him and handcuffed him and took him to jail. $30,000 fine and no more license in the state of Vermont. The last time I heard he was pumping gas."

Some young men and women start life with a map that whisks them through their education and into a career with no detours. And then there are those with no fixed plan but with energy, curiosity, and character who travel the byways until they hit the main road. Articulate, energetic, and eager to explain his work, Ron Mercier took the circuitous route but is happy where he ended up. He left Montpelier High School as a junior and over the years worked in construction, retail sales, and the restaurant business, earned his GED, and took courses at community colleges. He supervised the completion of the treatment plant in Bradford, decided that he had the skills to work there, passed the state test, and landed a job: "And boy, I stayed right with the books. The state got to know me and my character and what I knew." He started teaching the ninety-hour wastewater treatment course. Hired in 1981 to oversee the final stages of upgrading at Montpelier, he eventually was put in charge of the plant. He has been on the job ever since.

Proud of his accomplishments, interested in his work, he is, I said, a lucky man. "Yes," he said, "I think I am. The work is varied. I do computer work, I do drawings, I figure things out and set things up. I check with the guys to see what they're doing. And knowing that you're really accomplishing something, keeping the river clean. When I was a kid and used to fish out here, there was garbage and all sorts of things hanging from the tree limbs in the spring. And now when you look at it, it looks so much nicer, much better for everybody. I know it's not me directly, it's everybody, but you can see the changes, you can see something really valuable happening. I see it out here every day."

The more I learned about the valley, the more the discrete elements formed patterns, from interwoven streams to social relations among people. Everything had a dual character (or more): its indisputable uniqueness and its participation in a wider scheme of things. Montpelier's wastewater treatment plant was part of what Alice Outwater has called the "vascularization of a city," the network of conduits that brings water in and takes it out.[12] Montpelier's vascularization has done well the work it was designed to do, as have comparable systems in other towns up and down the valley. The watershed's fifteen municipal wastewater plants at peak permitted capacity could discharge 21 million gallons per day.[13] Even at their much lower actual rate, the impact on the river would be catastrophic if the waste were not treated as effectively as it is.

There is a wider problem, however, that the plants cannot address—nonpoint source pollution. Municipal sewage gathers at a single point near the Winooski and there it is treated. Substances such as agricultural fertilizers and pesticides, animal waste, lawn fertilizers and weed killers, oil and grease from automobiles and boats, a wide variety of other chemicals, household sewage in septic systems, and sediment from erosion—these are scattered throughout the valley. They are diffuse, and all can become pollutants of smaller streams and ponds, of the river, of Lake Champlain, and of groundwater if they are not properly controlled and contained. Nonpoint pollution is now the most serious source of contamination in the valley, the state, and the nation at large. When we think of oil spills, industrial or shipping accidents come to mind, but it is said that while refineries and factories were the chief culprits thirty years ago, today it is the individual who pours his used motor oil down the storm drain or on a field: "We dump an estimated 17 million gallons—more than the Valdez oil spill—into our watersheds *each day.*"[14] Nonpoint sources account for 90 percent of the pollution of Vermont's impaired waters.[15]

It is part of the human drama that we pay a price for every improvement. As a teacher of mine used to say, a straight line runs from the first arrowhead to the atomic bomb. A bemused god might see the play as a comedy, but such Olympian detachment is hard for most mortals to come by. Such is the story of the revolution in industrial and agricultural chemicals that brought us everything from prodigious food production to, it is believed, lower world-wide human sperm counts. Alice Outwater's packed chapter "Down the Drain, Up the Stack" is a revealing account of what has happened nationwide over the past fifty years.[16] Writing in 1996, she said that "there are some seventy thousand widely used synthetic chemicals, and about a thousand new chemicals are introduced each year." She recounts the disastrous effect of DDT on birds at the top of the food chain ("eggs encased in membranes instead of shells"); PCBs that deformed the young of mammals; the effect of various chemicals on the hormones of many creatures including humans (those lower sperm counts); and the astonishing pollution load of the Calumet River that in the 1960s received "a daily dose of about 100,000 pounds of oil, 35,000 pounds of ammonia, 3,500 pounds of phenols, and 3,000 pounds of cyanide from the dense industrial complex around Chicago, Gary, and Hammond." More obviously dramatic was the Cuyahoga River in Ohio that in 1969 was "so clogged with floating logs, oil, old tires, paints, and chemicals that . . . it actually ignited, burning two railroad bridges in a toxic inferno of flames some 200 feet high." Excessive phosphorus and other nutrients spurred the growth of algae that in Lake Erie "formed mats up to 2 feet thick, covering hundreds of miles of the lake's surface." And of course there was much else.

Fortunately the story doesn't end there. A growing environmental consciousness led to a number of federal laws including the Federal Water Pollution Control Act of 1972 (the Clean Water Act), which was to make all waters safe for swimming and fishing by 1983 and was to end the dumping of all pollutants by 1985. America's waters are much healthier now but by no means pristine, and the goals have not been met. One report estimates that as many as 40 percent of the nation's waters are still not safe for swimming or fishing and that even discounting certain violations of the Clean Water Act that the Environmental Protection Agency does not consider significant, "nearly 30% of major facilities examined (1,798) were in Significant Noncompliance (SNC) with their Clean Water Act permits for at least one quarter during the 15 months beginning January 1, 2000 and ending March 31, 2001."[17] Passing laws is one thing; effectively enforcing them is another.

The Winooski Valley and the state of Vermont at large share in the general distress despite great progress. Historically the Winooski has suffered many of the assaults inflicted on its bigger sisters elsewhere, although it has never caught fire. "By the early 1900s," according to the *Montpelier Rivers Report,* "nearly 30% of Vermont's manufacturers were located on the banks of the Winooski. The river offered power, cooling, cleaning and easy waste disposal."[18] It runs through the region of the state now most intensively developed and most rapidly developing but in the same lower valley has much rich farmland. Both urban and agricultural pollution are problems to contend with.

Phosphorus from Burlington's lawns and gardens is the biggest pollutant in the stormwater entering Lake Champlain's Burlington Bay.[19] Algae levels in parts of the lake sometimes spoil swimming, and in 1999 blue-green algae in another part of the lake produced a toxin that killed two dogs that drank the water. A controversy arose over run-off from a Lowes Home Center parking lot in South Burlington. Much of the valley is rural; failed septic systems continue to be a problem. And twenty-six reaches of the Winooski River, its tributaries, or lakes in its watershed have been called impaired by the Vermont Department of Environmental Conservation.[20] The state is required to take action, including perhaps establishing a Total Maximum Daily Load of each pollutant that can be added to the water. This assessment will be part of a comprehensive Winooski watershed plan when its time comes; plans eventually will be developed for all watersheds in the state. In Vermont's "Unified Watershed Assessment," the Winooski watershed is one of five with a "high need" for restoration and protection.[21] Like all state watersheds, it has some sensitive ecosystems and like all but one, it needs further assessment.

Given all this, it may come as a surprise that the biggest single pollutant of Vermont's waters and of the Winooski itself is none of the above but nature's own product, the soil or, to be a bit more technical, mineral sediments. As I made my way downriver in *The Tub,* I saw plenty of examples of streambank erosion, and later I was to see many degraded sections of the tributaries where massive amounts of sediment had washed into the stream as a result of human alteration of the stream environment. Some wash of sediment would occur if no humans had ever entered the valley—the fertile farmland along the river was in part the result of sediment deposits over the ages—but what I was looking at was far in excess of the relatively slow processes of nature itself. Eventually I was to learn that even more significant than the damage to the immediate location was the effect of the sediment on the aquatic habitat: embedding stream-bottom rocks, thereby shrinking the living quarters of aquatic insects in certain stages of their development, thus reducing food for the fish; clogging the gills of fish, making it more difficult for them to breathe; and covering the pebbles on the stream bottom where trout lay their eggs or smothering the eggs themselves.[22] This process, here abstractly described, was to become vivid and particular in the valley of the Mad River, a major tributary of the Winooski where I learned to see more clearly, thanks to the patient tutelage of the fisheries biologist Rich Kirn, the geologist George Springston, and the fluvial geomorphologist Lori Barg.

But first more adventures with *The Tub.*

11

Backyard Camping

The food alone weighs 29 pounds. —MY TRAVEL LOG

Here I am, back at the beginning of navigable water, for I canoed the river twice, first in day trips, now camping along the way. My camping preparations and gear are of safari proportions: two big duffels (clothes and food), tent, two tarps, sleeping bag, foam pad, rain gear, miscellaneous clothing, recorder, notebook, binoculars, Buck knife, multi-purpose tool, waterproof containers of matches, first-aid kit, and a variety of bottles, bags, boxes, cans, and lines. And of course *The Tub,* a kayak paddle, a back-up canoe paddle, an air pump, and a two-wheeled boat carrier for portages. *The Tub* loaded weighs well over 125 pounds, the food alone weighs 29 pounds. Arrangements are equally elaborate. I left home at 6:30 A.M. and drove the 127 miles to Les Snow and Lori Baker's farm in Marshfield where I left my canoe and gear, then down the valley to Burlington to the home of Azarius Beaudoin, my trapper friend on the shore of Lake Champlain, where I left my car. Louie of the Anywhere Taxi Company picked me up as planned and took me back to Les and Lori's farm, where I am about to embark.

The gear takes up half *The Tub* even when meticulously packed in a complicated contrivance of lines and tarps. I open the collapsible boat walker, pump up the tire that has developed a slow leak, then discover that, strong as I am, I cannot simply lift the loaded canoe from the ground to the walker. It is only now that the wisdom dawns upon me of mounting the canoe before packing. Embarrassed to start over—my antics are probably under surveillance from the farmhouse—I discover that by lifting one end onto the wheels I can crawl under and raise the canoe with my back while pulling the wheels into position. It turns out to be a useful skill, for as I set off across the bumpy pasture on my way to the river, the straps that came with the walker prove inadequate to hold it in place. *The Tub* twists on the carriage until canoe and wheels point in different directions. The brain is weaker than the back, for it will take several episodes of crawling, lifting, and shifting before it occurs to me to secure the walker with extra lines.

I finally lower *The Tub* down a steep bank into the river and set off, later than I have planned and worried that I might not make my first campsite by dark. It isn't far, but I have the portage around Plainfield village to contend with. I paddle faster than I would like and observe less. Slumping banks and point bars slide by at the periphery of my attention, although I do notice some mother ducks with their little ones scooting ahead of me and suddenly diving. I imagine a popping sound as they disappear. Thanks to the weight, the canoe rides closer to level. Emerging from the river a little before the village is even harder than getting to it. The bank is steep and slippery. Inch by inch I haul the loaded canoe out of the water, pulling until my muscles are about to tear. For effort that takes me right to the limit of my strength, this one exit is matched only by a portage later to be described. Long-term exertion, hours of continuous paddling, are another story, still ahead.

Curious eyes tell me that a man wheeling a canoe through Plainfield is a spectacle not often seen. Back in the river below the recreation field, where before I sometimes had to walk the canoe, the fast water now hurls me along at a great clip but slams me against rocks where I occasionally hang up. I wonder how much punishment *The Tub* can take, but despite all the abuse inflicted by this shallow, rocky river, her hull is to come through scarred but intact. Below the rapids, the river levels off and crosses Route 2 into East Montpelier. This is one of the most pleasant stretches, largely free of rocks and with no challenging turbulence but enough current to push me along. In high water, however, it is also an area of significant slumping and lateral movement as I noted in my first passage. Today my mind is on my campsite, and as the sun sinks lower I paddle faster until I spot the green junkyard fence that is my landmark. Just beyond I pull over to the bank, step out into deep mud, and fall flat on my back. Quicksand crosses my mind, but it isn't quicksand. I flounder to my feet and, smarter this time, unload the canoe item by item, carry them up the steep bank, pull up the canoe, reload it, curse the recalcitrant wheels, and with much puffing make my way around a hayfield to the backyard of Norm Ferland in East Montpelier village.

When I first thought of camping, I imagined wilder scenes, but it was the river that domesticated this valley, that brought humans to its banks for fish and good soil, for travel and power. For something approaching wilderness, you have to go into the mountains, although there are places where paddling along below the roads, between wooded and cliff-lined margins, I can imagine what the Abenaki would have seen hundreds of years ago. Now my tent is pitched behind Norm Ferland's barn, the result of a generous invitation extended long before when, scouting for campsites, I first met him. I sleep better than expected and just finish my breakfast of powdered milk, soy protein, a bagel, and dried fruit when Norm tries to give me a second meal. I accept coffee and toast, and I hear his story.

Norm was a high school music teacher at various schools in the valley. He is a man of strong feelings and opinions, of an ardent temperament and passionate interests, nearly all of them now frustrated. He loved music, teaching, his family, and fishing. In 1970, his second daughter, Lisa Marie, died at age eight of leukemia. The grief shook the marriage, and his wife eventually left him and married someone else. Three years before I camped on his land, he was operated on for a severe blockage of his carotid arteries. Lying in the recovery room, he saw a shower of flashing lights and was sure he was going to die. His right side was partially paralyzed: he had suffered a stroke. A lot of motion came back, he can walk, if somewhat unsteadily, but his right arm is virtually useless, the hand and wrist swollen and painful. He occasionally gropes for words, he can no longer play his beloved clarinet or the piano, he had to give up fishing, and he even gave away his huge black Lab who was too rambunctious to handle. He lives alone in an old house badly in need of paint and repair, apparently beyond his means to maintain. He keeps a photo of his daughter's gravestone on the refrigerator door.

But for all this he isn't sour. He loves life and is sad that so much of it is now closed to him. His critical judgment hasn't weakened. He is contemptuous of the permissive attitude that says "I'm okay, you're okay." Some people, he says, are not okay. He is too honest to pretend that he has not suffered great losses, and he is not alone in dreaming that if he could live his life over again, he would live it more boldly. Eager to have my company for as long as possible, he walks with me down to the river where I put in my canoe: "I would like to go with you. I would like the adventure of it. The spirit is willing, but the flesh is weak." I said that he had many good years of fishing. "Yes, but that's all in the past," he said, "not in the future." I seat myself in the canoe and pick up the paddle. Norm calls to me from the top of the bank: "I'm glad you're doing what you enjoy. Looking back, it seems to me all my life I was standing still."

I paddle on.

12

Stream Dynamics

The Water Woman and the Surficial Geologist

Any river is really the summation of the whole valley. To think of any river as nothing but water is to ignore the greater part of it.

—HAL BORLAND as quoted in the brochure, "Friends of the Mad River"

In the Roman West, from a very early date, rivers were conceived as roads: highways that could be made straight; that would carry traffic and, if necessary, armed men; that defined entrances and stations. The model for the well-behaved watercourse was the aqueduct: the highest achievement of Roman engineering. —SIMON SCHAMA, *Landscape and Memory*

A river without adequate riparian buffers is an unstable river system. Federal and state agencies, towns and landowners have spent millions of dollars trying to keep rivers from doing what rivers do naturally. Rivers are dynamic systems. A stable river will move through the landscape without causing damage. —LORI BARG, *The Streamside Sentinel*

Here's your mini-model of differential erosion, mountains and valleys illustrated in a single stone.

—GEORGE SPRINGSTON

When I was a boy, I was attracted to the Mad River by its gorges, deep pools, and romantic name, which spoke to me of something angry and untamed. Angry and untamed it sometimes becomes, despite the wise and foolish efforts of people to control it, but even at its quietist it claims pride of place in two respects among all the tributaries of the Winooski: it is the longest at 26 miles and it drains the largest watershed at 91,080 acres.[1]

It also has won the affection of an ambitious and effective volunteer organization, the Friends of the Mad River, which on its founding in 1990 took over the Mad River Watch Program, begun five years earlier, one of the oldest continuous volunteer water-quality monitoring efforts in the country. Although the River Watch is at the heart of its mission, the

Friends' ambitions are more far-reaching, as I was to learn from Michael Blazewicz, the Friends' then Watershed Coordinator.[2] Mike is a perfect exemplar of a rising generation of conservationists who see water-quality issues at the heart of broad ecological concerns but who want to blend conservation efforts with other local interests such as farming, housing, jobs, and a measure of community self-reliance. These conservationists are not indiscriminate tree huggers, but they believe that if some trees aren't hugged and others planted, swimming, fishing, boating, and farming are among the human activities that will suffer, not to mention the whole economy, which relies on natural capital. "Sustainable" is a key word.

Mike and his conservation compatriots are animated by a zeal not unlike that of religious converts, although the content of the vision is different. Not the heavenly city but an earthly community is the goal, a community in which the breach to be healed is not between man and God—man's first disobedience and his expulsion from the garden—but the breach between man and the natural world implicit in modern man's attempt to conquer nature for his own advantage. Mike and his conservation compatriots are by no means primitivists in any significant sense of the word. They are far better versed in the natural sciences and technology than the average American, and rather than avoiding science they draw on it and its accompanying technologies to do their work. But unlike Francis Bacon, one of the founders of the modern scientific world view, who wanted nature "under constraint and vexed; that is to say, when by art and the hand of man she is forced out of her natural state, and squeezed and moulded," they take a protective view of the natural world.[3] They use scientifically based technologies to preserve pockets of relatively untouched natural areas, and where human intrusion is unavoidable or even desirable, they try to integrate man and nature in a manner least destructive of the natural environment. Underlying most of the discussions and literature is the notion of balance, an equipoise of natural forces changing only slowly over time. Ironically, given the overwhelming success of our species, this natural balance or some simulacrum thereof can be restored only by the artful use of the same human powers that disturbed it in the first place.[4]

While a student in the University of Vermont's Environmental Program, Mike was given a copy of *The Best River Ever: A Conservation Plan to Protect and Restore Vermont's Beautiful Mad River Watershed.* Published in 1995 by the Friends, this is an ambitious, comprehensive plan nearly 200 pages long, meticulously organized, handsomely produced, with a description of various watershed problems such as erosion and sedimentation, lack of stream bank vegetation, pollution, threats to public access, and lack of information and education about the river—and recommended solutions.

Mike was impressed. Here was a group, he said, that's "got it together, doing grassroots environmental work on a watershed level." He struck up an acquaintance with Kinny Connell, one of the founders of the Friends, its president from the beginning and "a leader whose thought and caring are a great asset to the community." Eventually he secured a part-time job with the Friends working on various preservation projects. Of the most interest to me were the streambank stabilization efforts and the more comprehensive Watershed Assessment Program.

One of the most notable projects was the major experiment at the 1824 House Inn site in Waitsfield in the fall of 1996 and the following spring. This long, sweeping bend in the river prompted memories of lazy days: an open meadow right to the river's edge, a riffle, a deep pool, a generous outcropping of bedrock, and a huge point bar on the inside of the bend where a boy with a fishing rod reminded me of myself fifty years ago. But this happy scene was a disaster waiting to happen and happen it did with the heavy snows and rains of 1995 and 1996. The lovely meadow lacked the one natural thing that might have saved it, a sturdily wooded riparian edge. The high water ate away some twenty feet of bank, sending it downstream to add sediment to the already excessive load. The stabilization project was experimental in that along with other efforts in Vermont and New York, it revived "the techniques of an almost lost art called soil bioengineering. Soil bioengineering uses plants and organic structures rather than rock riprap to hold streambanks in place."[5]

With a grant from the Lake Champlain Basin Program and professional advice from the state's Agency of Natural Resources, the Friends and the Mad River Valley Planning District made a 300-foot-long revetment of fifty-five hemlock trees cabled to the bank, their tops pointing downstream, the purpose being to slow the water and capture sediment. Then they added brush rolls to cushion the bank from the abrasion of high water and ice. Next they stuck black willow posts and branches in the ground; planted in the fall, they would root and leaf out the following spring. The exposed soil was limed, fertilized, seeded, and mulched, and in the spring, a buffer strip 25 to 50 feet wide of shrubs and trees was planted at the meadow's edge. As I was told by one man who worked on more modest projects elsewhere, "This project is the model of what can be done if you throw everything at a site." According to Kinny Connell, fifty-six volunteers put in 430 hours of work.

Its fate? It withstood normal assaults with only minor erosion and, more impressively, survived the devastating flood of 1998 that caused enormous damage up and down the valley. But then, in an unkind turn, it gave way under the provocation of lesser storms, and the river encroached on the diminishing field. A river's behavior is never truly erratic, however, for it moves according to the complex variables of volume, velocity, stream shape, bed material, soil composition, gradient, sediment load, ice, and no doubt other factors. If it appears to be erratic, it is because we do not understand the intricacies of its working. The Friends speculate that over time the river had undercut the entire installation, causing it to fail.

Experiments are designed to teach us things, however, and failure is not a total loss. Another time one might repeat the bioengineering steps but add riprap (large boulders) to stabilize the toe or base of the slope at the very bottom of the river. Complete riprapping lengthwise and top to bottom might guarantee success, but such methods come with a price: they harm fish and wildlife habitat and they speed up the flow, causing problems downstream.

For twelve years, as Mike explained, the Friends worked on various restoration projects, but as more and more people wanted help, "We began to feel that we were slapping bandaids all over the watershed and not really getting to the root causes of the problems." It became apparent that you couldn't address the various sites as separate units: "If you try to fix something downstream, it becomes undone because of conditions upstream. Better to start from the top down, look at severely degraded streams, smaller streams, try to stabilize those before going into the main channel."

With this in mind, the Friends secured from the Agency of Natural Resources a 2002 Watershed Grant to assess watershed problems. In Phase 1 of the Agency's recently developed assessment method, Mike identified unstable stretches of the tributaries. I came onto the scene at Phase 2, joining a training session for volunteers who were about to collect instream data following a meticulously designed procedure. In Phase 3, sites selected for remediation will be analyzed in much more detail, "almost an engineer's perspective." Training the volunteers were two specialists whose disciplines underlie the assessment: Lori Barg, fluvial geomorphologist; and George Springston, surficial geologist. Thanks to them I was to get a closer look at how streams really work, the dynamics beneath the glittering surface and the deep, dark pools.

It was from Lori and George that I first heard the wonderful, mouthfilling term, "fluvial geomorphology." "Stream dynamics" serves as an informal substitute but misses the precision of the more intimidating term, which suggests that we are seeking the relationships among the flowing water (fluvial), the land over which the water flows (geo), and the form or shape of the stream, channel, and wider watercourse (morphology). If a year before someone had asked me what I knew about these things, I would have said that water runs down hill, and the steeper the hill the faster it runs. I would not have imagined the sophisticated precision with which all these things and many more are scrutinized, measured, and evaluated.

Phase 2 began on the sunny first day of June, and once again I congratulated myself for choosing a subject whose research called for poking about in streams. But this was a more systematic poking about than my earlier wanderings. With the data gathered, town planners, state and town road crews, and private landowners could make more informed decisions about river recreation, land use, and road and bridge construction. Lori said, for example, that one study indicated that undersized or improperly placed culverts cause about one-half of all the flood damage in the state.

We met at Folsom Brook, which joins the Mad River about 2 miles south of Waitsfield village. With the exception of a lovely waterfall and pool, there was not much to attract the uninformed observer (me), just the usual riffles, pools, a culvert, some riprap. A most ordinary stream, one of thousands in my hilly state.

We headed for the stream with assessment forms based on protocols developed by the Agency with the assistance of Lori Barg and others.[6] Assessment areas could be up to 2 miles long, but our practice stretch was only a few hundred feet. How much could you say about a few hundred feet? The answer, I found, was plenty.

I am always stirred by the ease with which the specialist addresses practical problems. For the rapid geomorphic assessment of the training site, we looked for channel degradation (how deeply the stream has lowered the elevation of the bed), channel aggradation (how much the channel has filled with sediment), over-widened channel (how much the stream has eroded its banks and perhaps become braided with interwoven channels), and change in planform (how much the channel, as if viewed from the air, has changed location and is characterized by mid-channel bars, islands, flood chutes where high water crosses meander bends, and other signs of instability). Each of these physical features was evaluated on a numerical scale. The volunteers practiced, numbers were compared, and Lori, George, and Mike offered their more expert opinions.

A new vocabulary came into play for me, among the terms the nick point (or headcut) for a sharp drop in a stream, perhaps caused by removing gravel, where the falling water, now with more energy because of the drop, falls, sweeps back, and scoops out a hole that then undercuts the streambed, causing progressive upstream erosion. This can happen quickly, our handout explained, for a mile of streambed has been known to be lowered by as much as 5 feet in a single storm. Niagara Falls is a giant nick point, having eaten its way upstream several miles over a geologically short time span. Exposed bedrock in the stream, which erodes very slowly, is called a bedrock control. The main deep channel is called the thalweg from the German for river bed. The wider channel is the area between the banks, often much wider than the normal flow. Channel avulsion is the cutting of a new channel and the abandoning of the old. Flood chutes are paths taken by high water outside the main channel, sometimes crossing meander bends, for example, leaving the meander as an oxbow familiar in aerial photos of many streams, including sections of the Winooski.

To speak more generally, what Lori and her colleagues were looking at was the overall pattern of a stream and how it might change over time. The slower the change, the better. Hyperactive streams are unhealthy, like the Mad River when it chewed up the field at the 1824 House Inn.

I had always thought of streams as simply flowing water. They are in fact water and sediment, with sediment the cutting teeth that give streams their power not only to erode sandy banks but over time to smooth or cut through rock. The balance between water and sediment is critical for stream stability.[7] Excess sediment or particle size causes build-up (aggradation). Sediment covers fish eggs and aquatic insects, increases water temperature because of the shallow water and heat-absorption capacity of the particles, and may even build up enough to block the smooth flow under bridges or through culverts. Increased stream volume or velocity, on the other hand, gouges the channel (degradation), washing excessive sediment downstream to cause problems elsewhere and perhaps giving the stream less access to its floodplain. Overflow onto a floodplain dissipates energy and reduces erosion.

Lack of upstream forest cover, new roads or buildings, and channel straightening are frequent causes of increased stream volume and velocity, which is why land use is a river issue and why our most aggressive methods of controlling a stream aren't necessarily the wisest. It wasn't immediately apparent to me why bends reduce velocity. It isn't primarily because the water bumps up against them; it is because the bends reduce the slope of the watercourse, and the water travels farther to drop from one level to another. Skiers who crisscross a slope take longer to descend than those who plunge straight down; a waterfall is the shortest possible distance between two elevations and produces maximum velocity.

I was surprised to learn how small a stimulus is needed to change the shape of a river, almost as if it were a living organism with sensitive nerves, a fish worm, say, that will twist and bend at the slightest touch. It is said that instability begins with the disturbance of as little as 2 percent of a watershed, that logging or development of 8 percent causes active reshaping, and that the most common consequence of these so-called hydrologic changes is the widening of a river, with many Vermont rivers now two or three times wider than before and therefore shallower and warmer. We were treated to a dramatic example of instability when we emerged from the mouth of Folsom Brook and saw, just upstream in the Mad River, a well casing, once underground where it should be, that now rose from the river bottom some twenty feet in the air and far from the bank that had given way in the 1998 flood.

All this and more feeds into the Rapid Geomorphic Assessment form, but volunteers also are asked to evaluate the stream as habitat and to record information about bridges and culverts. On the Rapid Habitat Assessment form, river conditions are viewed from the perspective of stream life from tiny microorganisms up to fish. Mike picked up a rock from the bottom of the stream. You could see a line marking how deeply it had been embedded in fine sediment. If gravel, cobble, and boulders are too deeply embedded, there are few niches to shelter tiny aquatic organisms and little clean gravel for trout spawning grounds. Moderate undercutting of banks provides shelter for fish, but slumping destroys it.

And then there are the bridges and culverts, often too small for the stream or installed at an angle to it. If too small, a bridge or culvert will speed up the stream by narrowing the channel, causing erosion downstream. Debris may build up at the entrance, forcing the water to dig at the banks around the barrier or to overflow the road. If put in at too sharp an angle to the stream, a culvert invites erosion and debris at the sharp bend. The Bridge/Culvert Assessment form, which explains all of this, asks volunteers to record many features of the structure including type, various measurements, presence of sediment, and angle of flow. I'll never look at a culvert or bridge the same way again.

Riparian zone planting is a step any streambank owner can take, and the benefits extend beyond protection of the owner's land. Farmers aren't the only ones to lose when their land washes into the river. In 1998, one flood in the Missisquoi River basin town of Montgomery dumped into the Trout River an estimated 8,600 cubic yards of earth containing some 6.9 metric tons of phosphorus, second only to the sediment itself as Vermont's most deleterious pollutant. The extra expense of dealing with phosphorus in farming, wastewater treatment, and run-off protection is estimated by the Agency of Natural Resources at $1.10 million per metric ton. The immediate, direct restoration costs to Montgomery came to about $120,000.[8]

There are other benefits to a healthy riparian zone. Insects drop from streamside vegetation to feed the fish. Falling leaves and twigs become food for the small organisms at the bottom of the food chain. For some amphibians, reptiles, birds, and mammals (mink, raccoon, otter, many others), the buffer area will be home; for others such as moose, deer, and bear it may be a travel corridor connecting critical habitat areas. Archaeological sites, remains of Indian and early European activities, may be saved. A riparian buffer can filter out much of the sediment and other nonpoint pollutants—according to one study as much as 80 percent of the phosphorus and nitrogen—that otherwise would have easy access to the water. Potassium, calcium, phosphorus, and nitrogen concentrate in significant amounts in the leaves of some deciduous trees, while nitrogen is mineralized by certain aerobic soil bacteria and fungi, and is converted to nitrogen gases by anaerobic denitrifying bacteria. The Agency of Natural Resources recommends 25 to 100 feet of buffer, but says that more or less might be appropriate depending on local circumstances.[9] It is most important to protect the headwaters. Plenty of professional advice is available from both private hydrologists and state and federal agencies. Some of the literature takes the do-it-yourself approach as, for example, the well-illustrated *Vermont Stream Bank Restoration Guide,* which outlines a step-by-step procedure and provides funding and human resources directories.[10]

Another day I went into the field with Lori Barg and Mike Blazewicz to study some tributary sites. This was still Phase 2 of the project but a more detailed analysis than that conducted by the volunteers. Among other activities, we stretched a tape measure from bank to bank and measured from the channel bottom to the tape at various intervals across the channel. This step was repeated along the stretch, the numbers contributing to a computer-generated profile of this section of the stream. Lori and Mike said that Phase 3, reserved for the most critical sites, would require many more exact measurements, the size of the substrate, for example, and cross-sections sometimes as far back as 500 feet from the main channel.

I learned that rivulets and great rivers obey the same physical laws, a small brook producing a sizeable delta where it joined the Mad River. Jogging the other morning after a night of rain, I found the phenomena of stream dynamics beautifully illustrated in the freshly washed cutouts along a gravel road: headcuts, eroding banks, multiple channels, graded sediment deposits, alluvial fans.

I also learned that hip boots are not always the footwear of choice; Lori and Mike wore sandals. Wading the Mad River, I went in over my boots and nearly fell. It was not only Mike, a tall, strong man one-third my age, but Lori too—sure-footed, medium frame, a little taller than my 5 feet 7 inches—who moved easily against the current while I struggled. Always determined to show the young folks I was still a man to be reckoned with, I reluctantly took her hand to help myself across. On shore my wet boots were too hard to remove, so I lay on my back and raised my legs to let the water drain down on me. Lori said it was not a sight often seen in the practice of fluvial geomorphology.

We stopped to admire a splendid bedrock gorge, white water rushing through a narrow channel. I heard in her exclamation the same blend of wonder and scientific interest that I found in Chuck Ratté and indeed most of the naturalists who favored me with their company.

I certainly found it in George Springston when I joined him and his student assistant Fiona Johnstone on a field trip in the Mad River Valley. The ultimate aim of his work was to arrive at a comprehensive understanding of the surficial geology of the watershed, in part because of its relationship to the action of the waterways. Like all fieldwork, it proceeded step by step in small increments. Exploring gravel deposits at an elevation of 1,900 feet left him as excited as a gold miner with shiny specks in his pan, for he had never seen them in Vermont this high before. They were probably kame terrace deposits, formed as water and sediment from melting ice and nearby hills gathered between the hills and the glacier, although the site may have been a delta in a high-level lake. George and Fiona divided the bank into units according to the seams of the deposit, then measured the seams and noted the particles in each.

At a lower elevation, we walked up a logging road to examine some sites. Bearded, tall, and lean, George was the very picture of a field geologist with an impressive assortment of professional gear in a pack around his waist. We approached a small exposed bank and scraped away the surface to reach what soil scientists call the parent material, the origin of the topsoil. We found a pebbly, cobbly, granular gravel. I discovered that I had been using these terms imprecisely. Pebbles are 2 to 64 millimeters in size, cobbles 64 to 256, and boulders 256 millimeters and up. Boulders are head size and above, Fiona added. The material had a loose texture; you could dig it with your hand. George matched its color with a patch in a booklet, each patch designated by numbers and letters, this patch labeled 10YR3/4. Three characteristics told us that this sediment was not simply laid down as the glacier melted (glacial till), but was deposited by running water. First, it was layered, and "till can have some layering in it but not this dramatic and pervasive," according to George. Second, although varied in grain size, "there would be a much greater variation if this were glacial till." And third, there was some imbrication, that is, the pebbles overlapped like fish scales, so were arranged by running water. All of this and the location as determined by the Global Positioning System were entered on a map and data sheet.

Next, joined by the State Geologist Larry Becker (successor to Chuck Ratté), we examined an old stream channel deposit, which contained boulders stacked up like tilted dominoes by high water in an obvious imbrication. The deposit had been exposed by a deep road cut. As George said, "the bulldozer giveth and the bulldozer taketh away." Here it gaveth, leading to considerable speculation and instruction. To demonstrate imbrication, George dropped small stones in the brook, which arranged them in an overlapping pattern. "It's not just the velocity," he said. "With the specific gravity of the water at 1.00 and that of the stones probably 2.5 to 2.7, the water supports almost half the weight." While disclaiming any expertise in macroinvertebrate biology, George looked at the underside of several rocks and pointed out sandy specks that were in fact little houses made by caddisfly larvae who secrete a substance that cements sand particles together.

Downstream, George pointed to some metamorphic layering in the bedrock and explained that although we think first of layers in sedimentary rock, severe compression can cause mineral grains to line up in a certain direction, so-called grain growth, giving rise to layering in metamorphic rock. Tight folding can cause slippage along the layers, and very fine faulting or shearing may also result. I learned that "faulting" is a "fine general term" for faults so big that continents shake, as along the San Andreas, or so small that only a microscope can detect them. George picked up a "common Mad River rock, muscovite mica, shiny on the top, and underneath some tiny black grains—magnetite—look how my magnet moves toward them. Here, I'll make you a present of it." On my desk as I write, it's about 2 by 3 inches, smoothed off by tumbling in the stream and by sediment-laden water, with a groove along one side, obviously a softer layer, perhaps just a place with a little less quartz: "Here's your mini-model of differential erosion, mountains and valleys illustrated in a single stone." I asked if it would have come from far away. Glaciers are always picking stuff up and dropping it again, I was told, but only a small percentage of the stones travel a great distance, most only a few miles, although now and then you may find one from as far as the Canadian shield north of the St. Lawrence. "But I did make a cool find yesterday," George said, "a beautiful piece of limestone that came over the mountains from the Champlain Valley."

The science of fluvial geomorphology at its most technical is baffling to the layman, but it lends itself to a general application that the novice can grasp, as in Lori and George's superb study of a major watershed calamity, "Floods and the Great Brook in Plainfield, Vermont." Great Brook, a tributary of the Winooski, was once a relatively stable stream, "teeming with trout, with beautiful deep pools for fishing and swimming."[11] Not that it was immune to floods, having caused significant damage in 1857, 1869, 1927, 1938, 1973, 1984, 1989, and 1990.[12] But as bad and expensive as the earlier damage was, a major shift occurred after the 1973 flood when, on the recommendation of federal agencies, the town removed much of the flood debris, bulldozed boulders and gravel out of the channel, straightened the stream, and riprapped extensive sections of bank. The brook responded as according to the nature of stream dynamics it was bound to do. It speeded up and both eroded and widened, and in subsequent floods did more damage than ever. I visited some of the mass failures and found myself calling them landslides. Lori and George say that these severely eroded banks range from 10 to 80 feet high and up to hundreds of feet in length. Some existed before 1973, but there are many more now, totaling along the lower section of the stream more than 3,400 feet.[13] You could once catch 5-pound brown trout in Great Brook, but now "it sounds like a fish story."[14] And from 1984 to June 2001, Plainfield's economic flood losses to roads, bridges, and private property along the stream amounted to almost $2,200,000.[15]

Less accessible to the layman but of great importance to the state's efforts to improve stream quality is the procedure outlined in *Fluvial Geomorphology: A Foundation for Watershed Protection, Management and Restoration.* Published by the Vermont Department of Environmental Conservation, this booklet states that fluvial geomorphology can put a "science based foundation" under the state's various water resource programs by providing "measurable, incontrovertible field data."[16] It is very helpful to know in advance how a stream reach will behave before more damage occurs. Over recent decades, researchers have developed a method of prediction based on the actual behavior of comparable reaches elsewhere in the region. Underlying all of this is the recognition that watersheds are interactive systems, requiring for their successful management a comprehensive understanding.

As I ventured further into the mysteries of fluvial geomorphology, I soon came to a humbling halt at a text of impenetrable sophistication. Although the table of contents of David Knighton's *Fluvial Forms and Processes: A New Perspective* is clear enough with chapter titles such as "Drainage Networks," "Fluvial Processes," and "Channel Changes through Time," the depth of analysis and the complexity of mathematical formulations left me gasping in awe and ignorance. I knew that raindrops can erode an unprotected surface, but Knighton cites studies that precisely measured the kinetic energy of rainfall and its compaction of the soil surface, producing increased runoff.[17] And this is just the beginning of the discussion. His book and the research on which it is based provide detailed quantitative explanations of the phenomena of stream dynamics whose general character all can understand once their eyes have been opened. On the subject of stream bank vegetation, for example, I read that "Smith (1976) found that bank sediment with a root volume of 16–18 per cent and a 5 cm root mat was afforded 20 000 times more protection from erosion than comparable sediment without vegetation."[18]

In my travels in the Winooski watershed, I found much to discourage a lover of rivers, but the more I learned about past and present, the more reasons I found for a guarded optimism. For all the shocks to river stability inflicted by human beings, the heavily forested landscape is much kinder to rivers than the cleared hillsides of a hundred years ago or even of my childhood at mid-century. State government, the Agency of Natural Resources in particular, understands the problems and opportunities more clearly than ever and is dedicated to watershed protection. Among the general public, there is an increasing environmental awareness that leads many people, farmers among them, to want to do what they can to minimize the impact of their activities; the various volunteer associations such as the Friends of the Mad River are but the most visible form of this awareness. And just coming into its own in Vermont is the comprehensive science of fluvial geomorphology, which promises to provide the technical grounding for conservation efforts that without it are less persuasive and less certain of success. Not the least of its virtues is that it can show how the long-term economic interests of landowners, towns, and the state often coincide with the habitat interests usually thought to be the exclusive concerns of hunters, fishermen, birders, and other environmentalists. In Vermont, at least, environmentalists have reason to hope.

With my brief experience of the practical application of fluvial geomorphology and with just a glimpse of this critically important science at its most technical, I will say farewell to the Mad River Valley (for now) and return to the simple pleasures and exertions of *The Tub.*

13

A Perilous Portage in a City of Rivers

Cars and trucks roar past me, barely missing the canoe. —MY TRAVEL LOG

Montpelier is beginning to honor its rivers. —HANS ESTRIN, *Montpelier Rivers Report*

After Norm Ferland's wistful farewell in East Montpelier, I paddle farther downstream, all the way to an old dam, following which a long, shallow, rocky stretch would make canoeing almost impossible. This is only the second day of canoe camping, but already I suspect that *The Tub* is carrying enough food for a company of voyageurs. I work up an aphorism: think carefully, choose conservatively, leave half at home. I wonder what the Indians had for gear when they canoed the Winooski. I look at myself—never mind the gear—and I see a narrow-brimmed hat, clip-on sunglasses, a life jacket, knee pads, sunscreen smeared all over my exposed parts. I'm a project all by myself, and then there's the front half of the canoe, packed tight. I have had a taste of loading and unloading, and I will come to dread portages and making and breaking camp.

I've discovered that I really don't have meals; I snack all day long. I'll try for more formal dining now, but first I have to get out of the river. The bank is very steep here just before the dam. I tie a line to a small tree and use it to haul myself and my bundles to the top, one by one, and then, with much grunting and swearing, *The Tub* itself. I sit down by my little tree and lay out an elegant lunch: a can of tuna fish, a bag of figs, a can of tomato juice. Sated, I put the wheels under the canoe and tie them fast, blow up the soft tire, load all the bundles, then take a closer look at the dam. It's 8 or 10 feet high with a foaming, churning pool at the foot, and I think of how I felt approaching it, a sudden urge to paddle straight on, imagining the exhilaration of cresting the top and plunging down, just as one imagines leaping off a cliff, soaring into nothingness. But I didn't.

A little farther on will be Winooski One's dam and hydro station and then more difficult water and the city of Montpelier. I could put in at the outskirts of the city, but the access is difficult, and getting out I would have to struggle up a steep bank bundle by bundle at a busy bridge before the amused gaze of a thousand people. I decide to portage to the junction of the Winooski and the North Branch, which I can reach by crossing the railroad trestle behind the Grand Union. It will be a long hike but easy going, I think, on the shoulder of the highway. Easy it is down the long hill into Montpelier, but just beyond the intersection where Route 302 goes left to Barre the shoulder shrinks until it nearly disappears. *The Tub* is too wide for the footpath on the south side of the road, it rides in the traveled lane of traffic, and soon I come to a blind curve. It is the evening rush hour. Cars and trucks roar past me at 50 miles per hour, barely missing the canoe. I can't go forward and the north side of the road has no path and no shoulder to speak of. I back up. What to do? I decide that the outside of a curve is better than the blind side and seizing a rare break in the traffic, I scoot across the road; one car has to slow for me. I have two fears. One is that *The Tub* and I will be spread across the pavement like jelly, the other that a cop will come along and do who knows what, incredulous that anyone could be so foolish. Again the canoe rides in the traveled lane, and I look neither left nor right, my face as blank as I can make it. It is like a bad dream, but in a few hundred feet around the bend I wake up to the sight of a sidewalk on the south side of the road. I wait for another break, scamper across. No racing current, no hidden rocks, no peril of land or water will threaten me as this blind curve has done.

Canoeing alone and no cell phone? Foolish, some people said, but instant communication was the last thing I wanted. The trip was not only a going toward; it was a getting away, getting away to see, to think, to enjoy the illusion of solitude. At a Montpelier gas station, however, I call home to let Eleanor know that I am, so to speak, still afloat. And now it is a short walk to downtown, across the bridge, and around the Grand Union to the trestle at the mouth of the North Branch.

Montpelier is a city of rivers, for here three tributaries—the Stevens River, the North Branch, and the Dog River—join the main stem, and the Winooski itself flows right through the city, often flooding it. Montpelier is also the state capital, the nation's smallest, its size a matter of contemptuous notice by an early traveler who wrote, "By that association of ideas which is so prominent a characteristic of the human mind, a little town, when the seat of government, will always impart its littleness to the legislature and to all its coadjutors."[1] Only a couple of blocks from the river sits the handsome capitol building where some of the laws affecting the waterways and wildlife of the valley are enacted (some are enacted in that other capitol, in Washington, D.C.), but my eyes are on the waterways and efforts of local people to protect and enhance them. I have found no better guide to the interaction between a Vermont city and its rivers than the *Montpelier Rivers Report,* a publication of the Montpelier

River Project of the city's Conservation Commission and a worthy companion of *The Best River Ever* of the Mad River Valley. Most of what follows I owe to this document.

Montpelier almost certainly attracted the Indians who would have been drawn to the main river and its tributary valleys for traveling and fishing, and to the fertile valley floor for planting their crops. It was later to attract white settlers for the same reasons and for water power. The soil, enriched by thousands of years of flooding, was exceptionally fertile for a state not widely blessed with good farmland. The glaciers that left a bony skin over the hills here cooperated with ancient lakes and streams to produce quite different results. Back when the retreating ice blocked the outlet of what is now Lake Champlain, so-called Lake Vermont rose up the valleys until what is now Montpelier was 150 feet under water. Silt and clay gradually settled out of the water, and later, when the lake retreated, the various rivers of the area added their sediments (alluvium) so that "Downtown, you would dig through forty to ninety feet of alluvium and clay before hitting bedrock."[2]

The Indians left a greater mark on the landscape than we used to imagine, but they did little or nothing to disturb the actions of the river, its need to flood and gradually to alter its channel. It wasn't long after white settlement, however, that building on the floodplain and clearing the land to the river's edge began to clash with the river's natural habits, and the now familiar history of river control—its successes, failures, and hidden costs—was underway.

From the time of the first white settlers in the late eighteenth century to 1992, twenty-five major floods have swept through Montpelier, the worst of course the great flood of 1927. In earnest but often misguided attempts to protect their property, citizens did as river neighbors did almost everywhere: they straightened channels, added riprap, and built up the floodplain until "the North Branch and the Winooski were converted from rivers that freely moved within their Montpelier floodplains to straight, uniform drains that were constricted by buildings, bridges, and concrete and rock banks."[3] In places one could be paddling through a canal. In the 1930s, the Army Corps of Engineers built flood control dams at Wrightsville, Waterbury, and East Barre. The Wrightsville dam on the North Branch, the "biggest flood control project in the history of the country" as of its time, was built by 3,000 men of the Civilian Conservation Corps in the two years 1933 to 1935, "using only shovels and wheelbarrows."[4] The dams reduced the incidence of flooding but changed the character of the streams in other less desirable ways, raising water temperatures, depriving the floodplain of nutrients from sediments, and increasing downstream erosion.

And of course there was the sewage, all of which went directly to the rivers until 1964 when the first treatment plant was built. In fact, here behind the Grand Union, the small dam I see on the North Branch was built in 1936 for a reason I never would have suspected. While not the sewers of Paris through which hunted men could run, the sewers of Montpelier did harbor rats, and the dam was built to flood the pipes to keep the rats from breeding there.[5]

It requires a certain poetry in the soul to rise to the notion of honoring rivers as the *Report* does, but the Montpelier River Project has practical goals in mind. A rivers overlay map in the *Report* shows places where greenways, parks, and trails could be established. The *Report* gives detailed instructions for riparian buffers, their width and management depending on the mix of intended purposes, bank stabilization requiring about 10 to 16 feet, for example, wildlife habitat and corridors 100 to 650 feet depending on the species. The streams would be safe for fishing and swimming. They are fishable now, although quality varies, but the *E. coli* counts suggest that they are not safe for swimming. Downtown river frontage would be made attractive and useable, grass and picnic areas replacing trash and parking lots. The present appeal of Montpelier—even now a fine place to live and work with its handsome buildings, beautiful natural surroundings, and the hum and buzz of state government—is much on the Conservation Commission's mind, for that which draws people is threatened by the number of people it draws, their need for housing, and the growth of industry and business. The underlying question animating the entire River Project is "How can we continue to improve the quality of life in Montpelier while accommodating its inevitable growth?"[6]

The answer to this question and, indeed, to the host of related questions that I heard asked up and down the valley doesn't lie in our ever-advancing knowledge of the natural world or in its accompanying technology. Or, rather, it doesn't lie there exclusively. In a democracy, the social bedrock on which the ambitions of the River Project and similar undertakings must rest is the opinion of the people, not the rapidly shifting public opinion sampled by the pollsters but settled convictions as to the right order of things. Win their hearts and minds and the rest will follow. What gives value to the many community action and educational programs described in the *Report* is not just the practical results, from lowered *E. coli* counts to trash removal, but the shaping of the public mind to absorb conservation principles as part of a way of life.

I paddle away from the city of rivers, find my campsite, pull *The Tub* up the bank, and set up my tent. It is ten after eight. There are wispy white clouds overhead, various shades of purple to the west where sunset is starting. Very romantic, I think, requiring a meal worthy of the setting. My food bags oppress me. If I could just eat more, I would feel better about them. I do my best with peanut butter and crackers, protein mix and powdered milk, vegetable juice, dates, a strawberry cobbler cereal bar, and a can of Spam. Spam on crackers is tolerable, minimally tolerable, preferable to starvation or cannibalism. I eat half a can, put the rest in my scrap bag. With dirty fingers I brush my teeth and floss. I choose the least bumpy place for my foam pad and sleeping bag. At 2:30 A.M. I wake up. A bird is singing. What bird sings at 2:30 in the morning? I change bumps and go back to sleep.

14

The Trout Fisherman

O Sir, doubt not but that *Angling* is an Art, and an Art worth your learning: the Question is rather whether you be capable of learning it? for *Angling* is somewhat like *Poetry,* men are to be born so: I mean, with inclinations to it, though both may be heightned by practice and experience: but he that hopes to be a good *Angler* must not onely bring an inquiring, searching, observing wit, but he must bring a large measure of hope and patience, and a love and propensity to the Art it self; but having once got and practis'd it, then doubt not but *Angling* will prove to be so pleasant, that it will prove like Vertue, a reward to it self.

—IZAAK WALTON, *The Compleat Angler*

There is [in dry fly fishing] an altogether splendid sense of escape and freedom which results from deliberately making a fool of oneself, and the man who can fish a dry fly with both deadly seriousness and the knowledge that his seriousness is ridiculously misplaced enjoys the full blessing of this relief. Accordingly, he comes to love dry fly fishing for what it is: an absurdity upon which he can focus his full powers of concentration with only superficial concern for the outcome. Whether a man raises a trout to his fly matters as much as it matters little, and as little as it matters much. It is because dry fly fishing proffers this comfortable option that it is both a field of battle and a blessed haven of peace.

—HAROLD F. BLAISDELL, *The Philosophical Fisherman*

The main stem [of the Winooski River] and the smaller rivers and streams which feed it present the local angler an opportunity to fish just about every type of moving water that Vermont has to offer.

—FARROW ALLEN, EDWARD ANTCZAK, and PETER SHEA, *Vermont Trout Streams*

The first fish I remember catching was not in the Winooski Valley but on my uncle's farm in Rutland where, under the watchful eye of my grandmother, I pulled out of East Creek a now-mysterious fish we called a red-finned salmon. I soon graduated to trout. My father was a decent fly fisherman who tied his own flies and fished the Winooski and many other streams in northern Vermont, sometimes with me accompanying him, never as patient and successful as he was but not exactly a novice either. But he is long gone and on my return to the valley it was my good fortune to spend a little time with Lawrence Pyne—naturalist, fisherman, hunter, former newspaper sportswriter, one-time editor of *Vermont Outdoors Magazine,* and as of this writing co-host of Vermont Public Television's *Outdoor Journal* and Northeast Regional Editor for *Field and Stream.* Although there are many fine anglers who practice the art in the Winooski Valley, I could not have chosen better than this accomplished outdoorsman.

Lawrence recalled someone saying that through straightening, "We have transformed our rivers into high-gradient sluiceways." Before European settlement, the river would have carried less sediment: "The Winooski is quick to muddy and rise—not good for dry-fly fishing, which is a visual sport."

On the other hand, we were then riding through the spine of the Green Mountains where small tributaries fed cold water into the Winooski. This alone would have made for pockets of good fishing, but Waterbury Dam on Little River lowered the temperature from its entrance to the Winooski shortly above Bolton Dam as far downstream as Jonesville or even Richmond. "The river is reborn at the bottom of the dam," Lawrence said. Bottom water some 80 to 90 feet down, probably 50 degrees even in summer, "acts as a gigantic spring for the Winooski River," one source says, cooling it in the summer, warming it in the winter.[1] (Little River itself, however, suffers from the periodic release through the power plant, eroding the channel, alternating with too little water, and at the time of this writing the reservoir has been drawn down for repairs to the dam and is not due to be refilled until 2006.[2])

Once again and with ever-increasing frequency, the tales my valley people told were converging into one master plot, the story of the intricate web of the natural world and human intervention.

"There's a good run," Lawrence said, pointing to an island with a current on either side and at the lower end a pool 10 or 12 feet deep where the currents converged. Trout will wait in a pool or behind a rock where the water is fairly quiet and wait for the riffle to bring them food and oxygenated water. They also will stay close to the bottom except when they rise to feed, for friction with the river bottom slows the current there. If they expend more energy than they take in, they are doomed.

We parked about half a mile below Bolton Dam and followed a cold mountain stream (63 degrees—Lawrence carried a thermometer) through a culvert under the railroad tracks to its delta at the Winooski. We came to a long, sweeping riffle that is one of Lawrence's favorite stretches of water and of many other people too, for it is heavily fished. This part of the river has some excellent trout water, "perhaps not the best in the state, but the best stream fishing in Chittenden County." A handsome, robust man, outfitted and equipped for the stream, a boyish enthusiasm in his eyes and smile, he was the perfect picture of the fisherman. His Orvis waders (they breathe but keep you dry), Orvis rod and reel, a vest with many pockets, and a cap with a band of flies over the visor were all similar to the high-end gear I saw on the obviously amateur couple on a previous trip who were slapping their lines on an unpromising stretch of water. "Gold Card anglers," Lawrence said, himself no amateur. "They walk into the Orvis shop with their Gold Card. Not knowing what they're doing. That's a prerequisite for Gold Card anglers." Orvis, a Vermont company, has been a maker of fine tackle for decades. The Orvis reel was one Lawrence's mother gave him on high school graduation. "You've taken good care of it," I said. "There are some things in life," he said, "that you should be able to give to people who will outlive you, and a good fly reel is one of them. You have to really abuse it to make it not useable."

Fine reels, Orvis and a number of others, are prized for use and as collectibles. If you had a Bogdan trout or salmon reel, for example, you would certainly want to pass it on. Made in a small New Hampshire shop by a father-and-son team, Stan and Steve, a Bogdan reel costs over $1,000 (I saw a collectible on the Internet that was going for more than $3,000). Among the distinguished users was one of Stan's former fishing companions, Benny Goodman. The main parts are precisely machined out of solid metal to produce maximum strength and durability (at a cost of over 90 percent waste). The result is said to be a virtually immortal reel with a smooth, even drag, the resistance offered when a fish pulls off the line in a run.[3]

Despite my early experience as a fisherman, I find the subtleties of fly fishing tackle almost as involved as the technical elaborations of fluvial geomorphology. When I was a boy, a fine bamboo rod was highly prized. In my inherited copy of *The Trout Fisherman's Bible* (1949), I read that "Perfection in the art of rod building has reached the ultimate in the Orvis impregnated split-bamboo rod . . . and in the Winston split bamboo," but later other materials including steel, fiberglass, boron, and Kevlar made a showing, and now I understand that graphite has won out.[4] Depending on the intended use, fishermen look to the action of a rod (where and how much it bends) and its length, typical lengths ranging from 7½ to 9 feet. The most basic purpose of the rod is obvious—to cast the fly to where a trout may be waiting—but there are other considerations too, such as providing the extension needed to tighten the line and set the hook, and the resilience needed to prevent the fish from snapping the leader.

The leader, now made of nylon, is the fine monofilament attached to the fly line. Its purpose is to carry the fly the last stage of the cast and land it with a minimum of disturbance on the water. It comes in varying lengths, level or tapered, and in varying strengths as indicated by the dead weight it will support, 2 pounds, 4 pounds, or whatever. Lines receive almost as much sophisticated attention as rods. They come in different weights, diameters, materials, colors, level or tapered (single or double-tapered), floating or sinking, and sinking at different rates. Tapered lines become narrower at one or both ends, the purpose as with the leader to land the fly delicately; a double taper allows you to reverse the line if one end becomes worn. For good performance you must match the line to the rod. Though important, the fly fishing reel is a relatively simple piece of equipment, compared to the more complex bait casting and spinning reels, since its main function is just to hold the line. In fly casting, you strip line off the reel with your free hand; it does not come off the reel on its own as in spin casting. With a short cast and a small fish on, you usually retrieve the line by hand, but you probably play a large trout by gradually reeling him in, and at that moment you want a reel that works smoothly and dependably.

Add a net to capture the tired fish at the end of the battle, hip boots or chest-high waders, and a jacket or vest to hold various items, and you have a basic outfit with the important exception of the dry and wet flies. But it doesn't take many trips to the stream to discover that you need a dozen other accessories as well: scissors, knife, nail clipper, fly and line dressing, bug repellent, tape measure, thermometer, sun glasses . . . you name it. And my description above tells no more about the technical ramifications of equipment choice than a car dealer's brochure tells about the various models of Volvo. An old fishing classic on my desk has eleven pages on leaders, their ratings of strength in pounds according to type of material and diameter, and a variety of knots including the blood, barrel, double surgeon's, wolf splice, and Flemish loop. All knots, I read in Jum Quick's *Trout Fishing and Trout Flies,* reduce the strength of the material, from a third to as much as four-fifths.[5]

Lawrence tied on two flies and we waded into the current. The lead fly, the one at the end of the leader, was an Irresistible made of clipped deer hair. Since it floats well, Lawrence used it as a strike indicator. The fly intended to take a fish was a Hare's Ear nymph tied to a tippet off the main leader and riding beneath the surface. A trout would, it was hoped, take the nymph in his mouth but would soon spit it out unless Lawrence set the hook, which he would know to do if the strike indicator disappeared even if he hadn't felt the strike. Later he attached a small orange ball, easier to see in the fading light, and remarked wryly that what bait fishermen call a bobber is elevated by fly fishermen, the aristocrats of angling, to a strike indicator. I have read an entire magazine column on the use of indicators, acknowledging that "some 'purists' still argue against their use," but describing several different kinds including yarn, pinch-on foam, and strike putty.[6] Lawrence admits the legitimacy of a wide variety of angling methods—indeed just last night I saw him spin fishing for shad on public television's *Outdoor Journal*—but the irony of his remark about bobbers will be lost if you don't understand that in the minds of some anglers there is a strict hierarchy with live bait fishing at the bottom, fly fishing at the top, and spin fishing with artificial lures somewhat ambiguously in the middle.

There are even more disreputable methods such as snagging, the illegal use of big treble hooks to snag, say, spawning walleyes in the Salmon Hole. Snagging fishermen are known as cigar guys. They would carry a lighted cigar or cigarette, and if a warden appeared, one touch with the hot end would melt the monofilament spinning line to send the incriminating tackle to the bottom. The snaggers forced the state to forbid all fishing in the Salmon Hole during the walleye run, depriving fly fishermen of the opportunity to fish for steelhead. "Some people don't give a damn," Lawrence said.

With no strikes on the Hare's Ear, Lawrence tried the flashier Zugbug, made of peacock herl, the strands of peacock feathers. I mentioned having bought a Plymouth Rock neck for my father to use in fly tying, and I was told that this breed of chicken was now raised primarily for this purpose, the Metz farm in Oregon being a world-renowned supplier. Still later Lawrence tried some dry flies, dry in that they float on the surface, and then a big nymph, the Martin Stone Fly. Anyone even slightly acquainted with fly fishing knows that these lures are but the tiniest selection of a mind-boggling array of wet and dry flies already in existence with new ones being created every day. More has probably been written about the choice and use of flies than about any other aspect of angling. With but a little reading in angling lore, you find that the use of artificial flies goes back into antiquity with a third-century A.D. description of fly fishing on the Astraeus River in Macedonia by the Greek writer Aelian; that a woman, Dame Juliana Berners, is said to have published in 1496 the first book on fly fishing to come down to us, *Treatyse of Fysshynge Wyth an Angle,* some of the fly patterns still in use we're told (there is uncertainty as to her authorship); and that with Izaak Walton and his successors angling treatises gathered in an ever-swelling stream, as it were, until they constituted a genre of their own with a literary history rich in romance as well as technical information.[7] "There is something quite wonderful about fishing those storied pools," says one writer about famous streams, "and one cannot help but feel reverence for the parade of famous rods that has fished before him," then naming legendary fishermen some of whom, like Gordon and Hendrickson, have given their names to trout flies.[8]

But how many artificial flies are there? I doubt that anyone knows. As far back as 1962 there were said to be about ten thousand, and now for some aficionados fly tying has gone beyond utility.[9] "I know some superb tiers who don't even fish," one writer says. "It's become an art form for them, and we see whole new shows devoted solely to fly tying."[10] All I can say is that their splendid variety speaks to me of the wonderful excess of life available to the human animal after the satisfaction of basic needs. Sport fishermen don't need the food to survive. If there is a need, it is for something colorful and elegant, something to please the eye and raise the spirits, a desire for form, precision, and play. It is the analogue to music and poetry for those so constituted to appreciate it. It is a poetry of sorts that I hear in the materials used in fly tying—red grizzly hackle, jungle cock eye, wood duck fan-wing feather, golden pheasant crest, gray mottled turkey quill—and that I find in the dazzling shape, color, and names of the finished products—Pale Watery Dun, Queen of Waters, Yellow Sally, Golden Demon, Bailey's Mossback, Alaska Mary Ann, Blue Fox, California Coachman.

As to which ones to use where and when, it takes a bolder man than I to prescribe. Some people believe in matching the hatch, that is, using a fly that closely resembles the aquatic insects emerging from the stream. Some students of fishing become students of entomology as well; the book jacket of Schwiebert's classic *Matching the Hatch* proclaims that "all identifications have been checked by entomologists of the United States Department of Agriculture." Some cast doubt on the conventional wisdom of imitating the hatching insects. The legendary fisherman Edward Ringwood Hewitt said that "If a fly is of the proper size for trout at the time the fishing is being done . . . its color and pattern are not of any great importance," although dark colors are preferred because they are easier for the trout to see against the light sky.[11] Some fishermen give detailed instructions how to trim a standard fly; Hewitt said that "Form of fly makes no difference."

If you followed Hewitt's advice, you could make do with very few specimens, and indeed something of the Hewitt outlook seems to be reflected in the more contemporary advice in David Guertin's web site, "Flies for Vermont trout streams."[12] Guertin suggests fourteen flies as a standard kit, a "rough guide," but then gives emergence charts for sixteen species of mayflies and six species of caddis flies, the artificial fly that most closely matches, the likely months and time of day of emerging (hatching), and the habitat (quiet water, riffles, or whatever). He warns that compared to those of Pennsylvania or the Catskills, Vermont hatches are unpredictable as is the response of the trout. On one day they will refuse to strike in the middle of a hatch; on another, "a hungry Vermont brookie will hit everything from rubber bands to tinfoil." His charts may still be of use, he says, even if "The cynical answer" to what fly to use "is whatever *wasn't* working yesterday. The slightly less cynical answer is that it all depends on a lot of things, almost all of which are completely out of your control."

Some hatches are productive but very short. On Caspian Lake in northern Vermont, for example, at about nine o'clock in the evening around July 4, you might consider yourself in paradise for ten minutes—"twenty if you're really lucky," says Lawrence Pyne—while the giant mayfly *Hexagenia limbata* emerges from the water.[13] This is the famous Hex hatch that tantalizes big rainbows and anglers alike, but the gathering darkness complicates matters. Lawrence's fishing partner Leighton Wass says, "I've seen grown men brought to tears trying to put a fine leader through the eye of a fly in dim light."

Some writers classify flies as imitators (matching particular insects or small fish), impressionists (giving the general impression of an insect without exactly duplicating it), and attractors (drawing the trout by bright colors without even resembling an insect), the most famous of this last, I've read, being the Royal Coachman, one I remember from my early fishing days.[14] Harold Blaisdell distinguishes between "imitative" and "provocative" fly fishing, the latter stirring in a fish "those intense emotions which result in indiscretion." In imitative fishing, he says, the important thing isn't the pattern of the fly but its behavior on the water, how nearly its drift resembles that of a real fly: "I'd much rather fish the wrong fly right than to fish the right fly wrong."[15] I won't presume to pass judgment on any of these delicate matters, only to say that almost every human pursuit prefers multiplicity to simplicity as if the mind isn't satisfied unless it can find or bring into existence phenomena commensurate with its powers of elaboration.

My presence at Lawrence's elbow probably did little for his concentration. He continued to cast and talk, now stopping to turn over stones to show me larvae of caddis and mayflies, the husk of a stone fly. Among the many insects eaten in various stages of development by trout, the caddis and the mayflies are most in evidence and probably the most important.[16] There are artificial flies to imitate both immature and adult phases. There are variations depending on species and there are many species, some two thousand of the mayfly with more waiting to be properly identified and named. The caddis larvae cement together little cases of sand and pebbles in which to live protected as George Springston had shown me on a tributary of the Mad River. The caddis enjoys a longer adult stage, living from two to eight weeks before mating and dying, and the female attaches her eggs to underwater plants or rocks. The mayfly exists in the water for months or years as a nymph, but lives in the air as an adult for only two hours to fourteen days, and the female drops her eggs above the water or just below the surface. For the amorous male mayfly the French expression for sexual consummation, *le petit mort,* is insufficient; for him it's *le grand mort,* since ecstasy leads to sudden death. The female soon follows once she has released thousands of fertilized eggs.

Camping out on Cedar Island in Lake Champlain when I was a boy, I awoke one morning to find the beach covered with dead insects that were probably mayflies. For truly stupendous hatches, however, I have heard of nothing to compare with the Mississippi River now that pollution has been controlled enough to allow the insects there to thrive (a different species of mayfly, adapted to the muddy river bottom). Both caddis and mayflies are sensitive to pollution and their numbers are a good quick indicator of water quality, "like the canary in the coal mine," one man said. Just a month ago I read of Mississippi hatches so heavy in McGregor, Iowa, that "they have to be shoveled from riverside streets and scraped from bridges with snowplows."[17] These hatches were reminiscent of the epic swarms of fifty or sixty years ago such as the one at Sterling, Illinois, on July 23, 1940, that heaped a bridge four feet deep with bodies.[18]

No matter how skilled the angler, success is not assured. Lawrence hooked a good rainbow—we saw the flash of red—perhaps 14 inches, but after a brief tussle it broke loose. The true fisherman is in it for more than the catch, however, and Lawrence spoke of the welcome solitude that is charmed into existence by the concentration and intimacy of fishing; other fishermen, the power lines, the interstate all disappear from consciousness. You see the landscape in a new way and more of it.

As we were leaving the stream, we bumped into Jim Carey, Lawrence's former boss as publisher of *The Burlington Free Press*. Stories were exchanged about the passion for fishing. Jim said that he was fishing here at night with two friends when a bat took a cast fly and bit the man who released it. Bawled out by his buddy when he turned on a light, he said, "But it bit me! I may have rabies!" to which the reply was, "Shut up and keep fishing. You've got twenty-four hours before you need to see a doctor."

Lawrence was the first Pyne in some time not to go to Princeton, where his father taught electrical engineering. Instead he majored in environmental science at Middlebury College and did some graduate work at the University of Vermont in natural resource management. Leaving school, he spent two years working, hunting, and fishing in Alaska, but grand as Alaska is for a man of his tastes and although he came from "away," as my parents would have said, he considers himself a Vermonter. "It's in here," he said, tapping his head. As for professional identity, "What I am essentially is a frustrated fisheries biologist."

To me he is the all-around sportsman, but it was with a District Fisheries Biologist that I rounded out my exploration of trout fishing in the valley. Then a sixteen-year veteran of the state's Department of Fish and Wildlife, Rich Kirn is the author of several meticulously documented studies of Vermont fish populations, but adroit as he is with technical analysis he is perhaps happiest out on the stream gathering evidence and observing habitat. I first joined him on Chase Brook, a cold-water stream descending from high in the mountains through the Sugarbush Ski Area to join the Mad River on the valley floor. Tall, lean, and sure-footed, he moved easily through pools and riffles, poking his electrofishing wand into every nook while I slipped and stumbled along behind him, asking questions and marveling as dozens of small brook trout floated or squirmed to the surface and into the waiting nets.

Electrofishing or electroshocking is a widely used technique for sampling fish populations. A 550-volt DC generator was placed on the bank of the brook near a culvert to which the ground cable was attached. In one hand Rich carried the positive, the so-called wand, at the end of a 250-foot cable, in the other a small net. Tim Appleton, a Fish and Wildlife Technician, handled the other net while Dennis Donahue, a Seasonal Technician on an externship from Paul Smiths College, fed out the cable as we moved upstream. The electrical field stuns the fish and draws them toward the wand; it can sting a man too if he puts his hand in the water. Rich and Tim put the netted trout in a bucket and later in a cage set in the stream.

Next the fish were transferred to a tub of water containing an anesthetic solution that rendered them inactive so they could be measured and weighed; a live, active trout is a slippery thing to handle. Very seldom is a fish killed by handling—all survived the day of my visit—but the anesthetic dose must be just right, 4 milliliters in 4 gallons of water. (Before World War II, methods were cruder, diverting the stream or using poison to kill all the fish so as to count them, and up until the 1950s in Vermont, fish management consisted mainly of stocking the waters and enforcing the law.) Rich measured and weighed the fish, gathering raw data for the reports he would later write. The fish were then put in fresh water, later back in the cage, and eventually released. So thorough is the process that a second run was made on the same stretch and the steps repeated for a much smaller catch with a total of about 250. If the second catch was large, above a certain ratio, it would suggest that a third run should be made. Fortunately for my aching muscles, on all three stretches that day Rich was satisfied with two.

Small streams like this one are abundant in the Winooski Valley and indeed up and down the state, and most of them have brook trout like the ones we sampled that day. Although decent numbers are found, "It's a pretty harsh environment," Rich said, "the spring runoff, the snow, the scoured stream bed, so the fish don't grow very big and don't live very long." The young of the year measure up to 2½ inches, yearlings up to 3, some three- or four-year-olds might go 7. In a big pool, big for this stream, we took one that came in at a surprising 11 inches. Under rare, ideal conditions, Vermont brook trout can top 20 inches and 5 pounds.

Although smaller than other stream trout, the rainbows and browns, brookies are the favorite fish of Vermont resident anglers, as surveys have shown, perhaps because they exist more widely around the state and many of us were introduced to fishing in the little brooks where they are found.[19] Perhaps too because of their extraordinary beauty, as these excerpts from an extended description suggest: a back "vermiculated with yellowish sinuous marks, against a jade and/or olive background"; sides "veritable kilns of burnished color, crimson and scintillating bronze and gold"; spots "red, lemon, and blue"; and lower fins "a cobweb-Dresden china combination of laminated porcelain, white, black, and basic pink."[20] They sure are pretty.

Also known as native or speckled trout, brookies are classified by some writers as char, different from true trout. Trout or char, it is one of only four salmonids (trout and salmon) historically native to Vermont and, with the lake trout, one of only two flourishing today, the arctic char or golden trout almost certainly gone and the Atlantic salmon making a very slow comeback under intensive management efforts along the Connecticut River.[21] One theory, described in Edward Migdalski's *Angler's Guide to the Fresh Water Sport Fishes,* holds that all trout were originally an Arctic fish that retreated before the glaciers and then separated in different environments as the ice melted and evolved into the many species we know today.[22] The aquatic biologist Rich Langdon says that if this in fact happened it would have been way before the most recent glaciation, for by that time the species we know were clearly established. Another theory, according to Migdalski, is that all salmonids were once anadromous, that is, salt-water fish that came up freshwater streams to spawn; some, it is argued, were cut off and adapted to freshwater for their entire life cycle.

The migration of fishes into Vermont from both east and west during the retreat of the last glacier is the current study of Rich Langdon and his colleagues.[23] Some fishes, especially those from the Midwest, were new species, colonizing for the first time; others were recolonizing, returning mainly from the east as conditions permitted. Brook trout, for example, returned from both east and west. All of them we now consider native, giving Vermont seventy-nine native species, more than any other New England state. (It's like the human population of the United States, for everyone's ancestors far enough back came from somewhere else.) Langdon says that it is commonly understood that any fish present at the time of European settlement is to be considered native or indigenous. Wild trout, to the biologist, are those that, whatever their origin, now reproduce naturally; there are no native brown or rainbow trout in Vermont, but there are many wild ones.[24] Despite the brook trout's official status as a native, however, there is some uncertainty, even unease, over its actual biological make-up, whether a true "heritage strain" exists. As fisheries biologist Leonard Gerardi has said, "I believe that there is a repository of brook trout natural heritage. But I couldn't say with any authority that they haven't been crossed with cultured trout," an opinion echoed by Rich Kirn who says, "We don't have a handle on the genetic make-up of wild trout and how much that has been altered over time."[25]

A cultured trout isn't one with refined tastes but one that has been raised in a fish hatchery and released in the wild. For some 150 years, millions of fish, either hatchery raised or transplanted from one body of water to another, have been stocked by the state and private individuals. Depending on one's interests, the results are beneficial, destructive, or indifferent. When brown trout first were imported from Germany in 1883 in the form of 80,000 fertilized eggs, many anglers were enraged, convinced that the brown would wipe out other trout, but eventually it became a highly prized game fish and, according to one writer, "the fish that really created the dry fly purist in America" because "it took the dry fly better than any of our native trout."[26] It was only eight years later that the first brown trout reached Vermont, stocked in Bennington County, although apparently it isn't clear whether Vermont's first browns were from Germany or Scotland.[27] Some time ago the two strains merged.

The Vermont Management Plan for Brook, Brown and Rainbow Trout (1993) states that stocking makes sense only where wild populations cannot make it on their own: "The use of cultured trout . . . should *not* be considered an alternative to the protection or restoration of suitable trout habitat."[28] The plan describes three distinct objectives in stocking: to establish a self-sustaining population in waters suitable for all stages of a trout's life; to supplement wild trout in waters that will sustain life for more than the immediate season but will not support a healthy long-term population ("put-grow-and-take stocking"); and to provide an "immediate harvest" in waters that will not support trout year-round or will not produce them fast enough to meet fishing pressure ("put-and-take stocking"). There is clear evidence that stocking browns and rainbows in some streams has displaced native brook trout, and some evidence that competition with stocked fish harms wild ones and that undesirable genetic changes may occur. Introduced disease is a possibility.

I grew up with a prejudice against fishing for stocked trout, at least before they had become acclimated and semi-wild, thinking that the odds were stacked against the fish, but the *Vermont Angler Survey* of 1991 indicated that between 40 and 50 percent of those surveyed considered put-and-take stocking "very important."[29] I am reminded of the dual mission of the Department of Fish and Wildlife, almost inevitable in a democratic society and important also in the management of the deer herd, and that is to work in the interests of ecological well-being while at the same time responding to the wishes of the people—sometimes a delicate balancing act. (The Vermont Constitution confers the right to hunt and fish.) From an ecological perspective, the money and effort that go into stocking might be better spent on habitat improvement, but stocking is certainly here to stay. There is another reason for it, one that indirectly benefits native trout and all the programs of the department: fishing (and hunting) license fees are a significant part of the department's budget. People won't buy licenses unless they think they have a reasonable chance of catching more than a cold.

My second excursion with Rich Kirn took me not to a mountain stream but to the Dog River, a major tributary of the Winooski flowing 18 miles north from Roxbury to the main stem at Montpelier and draining 94 square miles. Stocking was discontinued in 1992 after Rich's studies showed that there were healthy populations of wild trout and poor catches of stocked fish. "I have seen more browns over the 18-inch mark on this stream than on any other I have visited," says guide and writer Peter Camman, and indeed the Dog is widely acknowledged to be one of Vermont's best trout streams.[30] In the *Montpelier Rivers Report,* Rich Langdon attributes the great productivity of the Dog to the diverse habitat: the riffles that carry food, the deep pools that give rest and a waiting place, the undercut yet relatively stable banks that provide cover.[31] The *Report* explains that calcium carbonate dissolved from the bedrock gives the food chain a boost, encouraging algae, fungi, bacteria, all feasts for insects. The relative hardness of the water—dissolved minerals at 54 parts per million compared to the less productive North Branch's 24—provides trout the minerals they need to grow large.[32] Well-shaded and fed by springs upstream, the river runs cold, carrying the dissolved oxygen needed in high concentrations.

Here on the Dog the generator was carried in a canoe, the ground was a 1-by-3-foot piece of sheet metal stapled to a piece of plywood trailing in the water, and there was a crew of seven, three of whom worked the wands. My modest contribution was handling one of the cables. The banks were well vegetated, although there were unfortunate stretches of the invasive, shallow-rooted Japanese knotweed, crowding out native plants better equipped to hold the soil. (This is the plant I know as bamboo with which I fight a yearly battle.) I happened to be following Tim Appleton, who amazed me by spotting in a swarm of tiny fish the one or two miniature trout and netting them. He said it was partly the behavior; the trout tend to come more directly to the wand. The other species noted, but not weighed or measured, included sculpin, black-nosed and long-nosed dace, suckers, lake chub, and creek chub. In the small mountain stream of my first visit, only brook trout were found.

The trout here were all rainbows and browns, and they were fewer and smaller than in recent years, the largest about 14 inches. Given natural variations, one year's tally tells little; if a trend develops, an investigation might be in order. Rich Kirn said that occasionally brook trout are found in this stretch, but their best habitat is upstream in colder water. All trout like cold, highly oxygenated water, but there are differences among species. While standing much colder temperatures, all do best from the mid-50s to mid 60s Farenheit. Just a few hours at a temperature above 75 degrees Farenheit usually will kill brook trout, browns can stand 80 degrees for a short time, and some strains of rainbows can go a bit higher. There are other differences. Brookies prefer an equal distribution of riffles and pools, they feed mainly in the daytime, and they spawn in the fall. Brown trout like pools to dominate, mature fish in streams tend to feed at night, and they too spawn in the fall with eggs, like those of the brookies, incubating over the winter. Rainbows are more sensitive to acidity than either brook trout or browns (acid water shortens the life of their sperm), but they often overlap with brook trout habitat upstream and brown trout habitat downstream. This pattern is found in the Dog River. Unlike the other two trout, rainbows spawn in the spring, although some hatchery rainbows have been bred to spawn in the fall. Rainbows like to spawn in streams; steelhead are lake-dwelling rainbows that swim upstream to spawn. Lake-dwelling browns also will travel considerable distances to spawn upstream. In all three species, the female hollows out one or more nests or redds in the gravel stream bottom with her tail and deposits her eggs, to which the male adds his milt. Excessive sediment can suffocate the eggs.

Browns grow larger than the other two, followed by rainbows. On my study wall is a mounted 24 incher, 4¾ pounds, caught by my father in September 1965, with the Mepps spinner still in his mouth (my father, though a fly fisherman, was not a purist), but they grow bigger. As of 2001, the all-time state record for brookies was 5 pounds 12 ounces, for rainbows 13 pounds 8 ounces, and for browns 22 pounds 3 ounces. All were dwarfed by the record lake trout of 34 pounds even.

Lest my emphasis on trout should mislead, let me say that the Winooski River is home to about fifty species of fish.[33] As a boy I caught pumpkinseed, rock bass, yellow perch, and smallmouth bass, and was familiar with dace, shiners, suckers, and a few more. I didn't know that there were eight species of shiners, and I had never heard of the banded killfish, the tessellated darter, the fallfish, or a number of others. Garnering considerable interest today is the lake sturgeon, an ancient species sometimes living 150 years and reaching 7 feet and 300 pounds, although none approaching this size have been seen in the Champlain basin for many years.[34] Low numbers have worried biologists for years, but in 2003 they found eggs and larvae in the Winooski below the Salmon Hole. Females don't spawn until they are about twenty-five years old and then only every five to nine years.

I have five of Rich Kirn's trout evaluations on my desk. There are also angler interviews to determine for various stretches the results per unit of fishing effort (hours spent), the species and size caught, the ratio of wild to cultured fish, and the impact of different kinds of tackle (bait, lure, fly). Rich's data enable him to estimate not only the number of fish per mile of stream but also the biomass or population in terms of pounds per acre. On one section of the Dog River, for example, he estimated 1.2 pounds of brook trout per acre, 15.2 pounds of brown trout, and 43.6 pounds of rainbows. The figures are meaningful because the size of the trout, as well as their numbers, is a measure of habitat productivity. Five hundred trout with an average weight of 5 ounces are not eating as well as 500 of the same age with an average weight of 10 ounces. With enough data of this sort accumulated over the years, the Fish and Wildlife Department can make adjustments in regulations governing tackle and the number and size of captured fish for various stretches of streams. (Rich's reports are aimed mainly at fellow professionals; the general reader and angler will find much practical advice in Camman's *Fishing Vermont's Streams and Lakes;* Hickoff and Plumley's *Flyfisher's Guide to Northern New England: Vermont, New Hampshire, and Maine;* Blaisdell's *The Philosophical Fisherman;* and Allen, Antczak, and Shea's *Vermont Trout Streams.*)

A study of the Sugarbush Ski Area for 1987 to 1996 showed improvement in habitat protection.[35] Rich confirmed what I had heard during my inquiries into stream dynamics and deer hunting: "It's piecemeal development that really harms us. Act 250 is pretty effective with developments. It's a house here, a house there, a culvert here—usually too small, not put in right." It's ironic that many of the people who build on their few acres back in the woods consider themselves faithful environmentalists. When they think of environmental damage, they think of chemicals and sewage. Their hearts are right.

The master document guiding the Fish and Wildlife Department's efforts on behalf of trout fishing is *The Vermont Management Plan for Brook, Brown and Rainbow Trout (1993),* mentioned above. Like the department's deer management plan, it evokes an ambivalent response. I have never entirely outgrown my boyish imaginings of a nature independent of man. The language of management, productivity, and harvest is imported from an alien realm. That said, however, a more rational, responsible part of my mind tells me that given the world as it is, the only hope for some semblance of a balanced natural order rests with the knowledge and dedication of professional managers, supported by an informed public.

15

The Long March and a Midnight Alarm

Take out 15 minutes or so for a snack—chocolate milk and peanut butter crackers in Waterbury. About 5½ hours. 20.3 miles.

All that remained was a plastic bottle (my bailer), my old sneakers, and a piece of terry cloth I had used as a kneeling pad.
—MY TRAVEL LOG

I rested at the gorge above the Middlesex dam where cliffs rose sharply out of the water, originally "fine sandstone and shale metamorphosed into a dirty quartzite with a lot of mud," as Chuck Ratté had explained, the dark bands probably volcanic dust that settled in shallow seas and formed a greenstone of chlorite, hornblende, and epidote. This was geologic poetry, dimly understood, but it was clear that great force had been required to bend horizontal layers into these nearly vertical folds. As I ate my peanut butter sandwich, food of the gods, my eyes rose to the majesty of the cliffs and descended to the ubiquitous plastic bottles and dark brown scum gathered behind the orange barrels strung across the river to warn canoeists of the dam.

This morning I came to a scene right out of my boyhood: a pasture gently sloping to the river, cow paths deeply etching the sandy soil, the sun beating down, and, in the river, several young cattle swishing their tails, splashing a little, cooling off. For a moment I was a boy again, and I almost wished that I had never learned that bare banks erode and do not filter, that cattle stir up sediment, that their manure carries pathogens.

I put in below the Middlesex dam and now, an hour later, I have stopped for a drink, a pre-emptive strike for I'm not thirsty. This is a dramatic stretch of the Winooski. I am sitting on a smooth, sculpted outcropping and have just passed high north-bank cliffs with cedars, birches, and blossoming columbine somehow finding toeholds in the steep rock face. I want to call it a gorge, but a state publication says that to qualify as a gorge a stream channel must have "continuous rock walls at least 10 feet high on both sides," and the south side doesn't measure up.[1]

A little while ago I saw a leafy branch about a foot long proceed in a steady, stately manner across the current. Curious about something so contrary to the laws of physics, I approached cautiously and saw a small brown body beneath the branch, kicking steadily. When I got within about 10 feet, the muskrat and the branch made port in a hole in the north bank.

I have come through a challenging set of rapids but not with all my colors flying. I didn't portage or guide *The Tub* with lines from the bank—that much I can say for myself—but I hesitated between the middle passage with its threatening rock and a safer path closer to the north bank. I was about to risk the more difficult route to prove my manhood when a boy and man came along to fish a few yards downstream. That decided it. The humiliation of capsizing in front of them was too grim to contemplate. I cut across the current to the safer passage, and emerged upright but crestfallen.

Not far below the dam I passed the mouth of the Mad River, which flows in from the south. I like names, even lists of names as in the old epics, even names that get attached by happenstance with no regard to the essential character of the thing named. Happenstance couldn't be said of the Mad River, of course. But the Stevens Branch, Ralph Nading Hill tells us, was named for a trapper who, disappointed in love, left his carpenter trade in Corinth, took to the woods, and was found much later lying dead next to the stream.[2] Nearby was his camp, and "Over the charred bed of an old fire hung a camp kettle with water and herbs for sickness in it." The Dog River, which should have been named for its trout, was so called because "a pioneer hunter's dog got caught in a beaver trap and drowned in the river trying to free himself." As for the Kingsbury Branch, "A pioneer named Kingsbury fished in it a great deal of the time and always referred to it as *his* branch. His neighbors agreed." And back in Cabot there was Molly's Pond, one of the sources of the Winooski, with its neighbor Joe's Pond, named for Indian lovers about whom various stories are told. I now live in the small sub-watershed of Salmon Brook in southern Vermont, and there are a number of other streams in Vermont named for their natural characteristics, but a great many are identified by some human association, for we appropriate the landscape by speech as well as by deeds. In a list of twenty-one tributaries of the Mad River, for example, four are unnamed, two are named for natural features, and fifteen bear the names of human beings or human activities.[3] I wonder how Sodom Pond Brook in East Montpelier got its name.

Mention of the unfortunate Stevens and other early pioneers brings to mind the experience of James Marsh, Waterbury's first settler, "the greatest hard-luck story," Hill says, "in colonizing the Onion."[4] Losing a farm in New Hampshire because of title problems, Marsh came to Waterbury, cleared land, and raised some corn, and eventually brought his family to join him. A series of hardships followed: children left alone running out of food in their father's absence, work downstream for which he wasn't paid, the family living on wild onions and milk, carrying a bushel of corn on foot from New Hampshire, a flood washing away the corn crop, hunting moose to swap the meat for corn. At one point, however, hungry as he was, he wouldn't eat his wife's wild-onion goulash. After all his heroic efforts, James Marsh fell through the ice and drowned, and a heavy snowfall forced his family to snowshoe 15 miles to attend his funeral.

Marsh's exertions were matched by many of the first settlers in northern New England. Samuel Pettengill tells of one Seth Hubbell who settled north of the Winooski Valley in Wolcott and left a written account of his trials. What strikes me first is the sheer muscle power needed to survive. For three years he carried a bushel of grain 12 miles on his back once a week to be milled. Once he carried a moose hide 30 miles to exchange for a bushel of corn. There is said to have been a common saying that if you couldn't carry 100 pounds 10 miles, you couldn't survive as the first settler. At first Hubbell had no farming tools except a hoe and an axe; there were no roads, and after a while no oxen to travel with or to plow. It was not uncommon for oxen and other cattle to starve to death as Hubbell's did, for where we now see grasses and grains growing in valley meadows, at first there was almost nothing but forest. Indians grew corn and pumpkins in a few places, but of feed for livestock there was only low-value "'browse,' a mixture of wild grass, leaves, twigs, sedge, ferns and small brush," around beaver ponds and in scattered clearings. Nearly all the nutritious grasses and grains that were to make Vermont an agricultural state were imported from abroad. Vermont was part of "the greatest continuous forest of the western hemisphere," stretching from the Atlantic to the Mississippi, so dense, it was said, that "a squirrel could go all the way from Plymouth Rock to the Mississippi River without touching ground, eating nuts as he leaped from one tree to the next."[5] (This expanse was punctuated, we now know, by Indian cropland, although less so in Vermont than farther south.)

As I leave my sculpted rock rest, I paddle through trout water good enough to win wide praise for its wild and stocked browns and rainbows, good enough for one veteran guide and angler to say, "I have caught several fish in there over four pounds and that was before the trophy trout stocking started."[6] The fish I don't see. What I do see, especially when I stop and climb the banks to look around, is that I am penetrating deeper into the Green Mountains. As the topo map's thickly clustered brown elevation lines indicate, here the mountains get steeper and higher. There is still a valley floor with fertile fields, but from Montpelier down there are places where the interstate, Route 2, the Central Vermont Railroad, and the river share strips of land as little as 500 feet wide. The Winooski is the grandfather of all these transportation routes, for it was used for thousands of years by the Indians.

At Little River I had come 8 miles from Middlesex and now, a mile and a half farther, I have reached the Bolton falls dam. The portage here is 2,000 feet, an eight-minute walk on a level path. But this is not a level path; it goes sharply up, flattens briefly, and goes sharply down. I make two trips, one with *The Tub* on my shoulders, one with paddles and other gear. Now and then, steadying *The Tub* with my back, I lean one end against a tree, the other on the ground, while I rest my legs. Three mountain bikers hurtle past me down the path, hares to my tortoise except that wherever they're going, they will get there first. The portage ends at a picnic area next to a wide stretch of river with a clear view of the water plunging over the dam. I push off. It's a luxury to be afloat again.

On the second leg of the day's journey, here in the very heart of the Green Mountains, I think again of the persistent force of the river, older than the hills, cutting down as the land rose up, its success still something of a puzzle to geologists. At the village of Jonesville, I pass the mouth of the Huntington River on whose banks sits the farm of my friends the Moultroups, and then I come to the bridge in Richmond and pull out of the river just beyond. The topo map shows how very quickly the sharp, high ridges of the mountains give way here to foothills and a broader floodplain, features seen less clearly from *The Tub*. The map's contour lines and colors signify more clearly the more I study them. I begin to see the flat paper three-dimensionally; I might be in an airplane, looking down with a touch of vertigo.

I drag *The Tub* to the top of the bank and find that I'm at Volunteer Green, the Richmond recreation field. With little cover to hide my canoe, I chain it to a sapling in plain view, hoping that public notice will protect it. It's the day for a farmers' market, so I look around a bit, then begin the hike back to Middlesex. For some reason I choose Route 2 instead of the quieter gravel road on the south side of the river and, as night falls, discover that here as in Montpelier walking can be more dangerous than paddling: narrow road shoulders, dull clothing, and for warning only a blank page of my log exposed to the lights of oncoming cars. A brief stop in Waterbury for a snack, and I move on. After five and a half hours of walking, I reach my car at Middlesex Dam; it is 10:35 P.M. Driving back to Richmond by the same route, I find that I have trudged just over 20 miles.

I look forward to telling people about my feat of endurance and I know my silly pride will catch their attention, so soft have we become. Pettengill mentions one Poultney settler who "walked thirty miles to Manchester with one hundred pounds of iron on his back, and returned the next day with one hundred pounds of salt."[7] With just three partridges and one regular meal under their belts, Ira Allen and four companions walked 70 miles through the woods in four days. Surrounded and supported by kinds and amounts of energy unimaginable to my ancestors, I seldom reflect on what it meant to do without them.

> Aside from their own muscle power, the only other sources of energy available to pioneers were the magnetic North Pole which gave direction to a compass; the chemical energy of gunpowder; the thermal energy of burning wood in the fireplace; the energy of gravity which made it easier to move men and cargoes down stream, and harder up stream or up hill; the energy of wind on the sails of ships, and the energy of the sun, of course, which causes the sap to rise and all living things to grow. I have read of only two windmills in early New England. Without energy, nothing is accomplished, even thinking.[8]

For all of that, degenerate modern that I am, I drive back to Richmond with crest high; I have redeemed my timid avoidance of the dangerous channel.

At Volunteer Green, a blow! *The Tub* is gone; gone my life jacket, gone my paddles; gone my waterproof bag with topo maps, sunscreen, small tools; gone my cocky assurance. A plastic bottle, sneakers, a piece of terry cloth, and a bent sapling are the only signs that I was ever there. The canoe must be in one of two places: behind someone's garage if an adult stole it, floating down the river on its own if kids took it. I can imagine the pleasure in their wicked hearts as they set it adrift. I go to the police station, but it's locked tighter than a jail. I call home from a convenience store; Eleanor is very distressed and sorry for me. I call the police and soon a young policewoman appears. Attentive and methodical, she inspects the scene very carefully: "Sometimes crooks leave things behind, like their wallets." But these crooks were not that careless. She is about to make some notes when a man appears on a bicycle. Very odd, I think, to be visited at midnight by a man on a bicycle, but he turns out to be the good angel. From his house across the way, he saw some youngsters dragging a canoe around the field. He rescued it along with all the gear and stored everything safely in his garage. A volunteer fireman, he heard the police report on his scanner. I go to his house and sure enough, there it is, everything, safe and sound. We talk awhile and exploring the social web of this tight little valley, I discover that when for a brief time my father was a paid fireman at Fort Ethan Allen in Colchester, his father was chief there and later chief in Burlington. I offer him a reward, but he refuses: "I'm just glad it's in the hands of the rightful owner." His name is Harland Stockwell, and my hat is off to him. From being a den of thieves, Richmond becomes a town of noble citizens.

I reach home at 3:30 in the morning, happier than if *The Tub* had never left my sight.

16

The Farmers

The study of grasses would be of great consequence to a northerly and grazing kingdom. The botanist that could improve the sward of the district where he lived would be a useful member of society: to raise a thick turf on a naked soil would be worth volumes of systematic knowledge; and he would be the best commonwealth's man that could occasion the growth of "two blades of grass where only one was seen before."

—GILBERT WHITE, *Natural History of Selborne and Observations on Nature*

The Moultroup farm in Richmond straddles the Huntington River, the stream that both helped to create it and now gnaws at its flanks as if to devour its offspring. Like its sister tributary the Mad River, the Huntington is a changeable stream, a sprightly, 20-mile good neighbor most of the time but a rude assailant in high water. The causes are of course the same, not really a deranged personality but a combination of geology and the physics of moving water in a landscape radically altered by humans. All the causes of stream instability that I have been taught since to recognize elsewhere were in evidence in the Huntington watershed as indeed in virtually every watershed of any size throughout Vermont. It was in fact an attempt by Jeff and Debbie Moultroup to restore some stability that drew me to their farm in the first place. I came to study an application of bioengineering, but I stayed to become friends with a family and its farm.

In 2001, Jeff and Debbie received the Conservation Farmer of the Year Award from the Winooski Natural Resources Conservation District. When I first met them, in 1997, their project was just underway. It was in effect two projects, combining streambank control with intensive grazing farming, both of which were new to me at the time. On my first visit, Dave Braun and Bob Magoon of the Winooski Watershed Association with workers from the Vermont Youth Conservation Corps were busy anchoring small hemlock trees to the river bank, a tree revetment similar to that at the 1824 House Inn on the Mad River, only on a much smaller scale. It was hoped that the trees, laid butt end upstream and cabled to the bank, would slow down the current and catch sediment at a long bend where for years the current had been chewing away the land.

Although often successful, bioengineering techniques do not guarantee success. Rivers want to move, and the more they and the landscape are disturbed the more violently they do so. The problem is not the strength of cable and rod—they will hold 3,000 pounds—it is the scouring force of current and ice that can undermine the revetment and cut behind it. And so it was that when I revisited the Moultroups' site several years later, the hemlocks were gone, the line of the former stream bank was now in the middle of the river, and 50 feet of land had washed downstream. In the meantime, a new technique had been tried, three vanes of large rocks angled upstream into the channel and sloping down into the water. "Counter-intuitive," as one writer says, "They look as if they'll scoop the water up and against the bank but just the opposite happens. The water, moving up toward the bank, rolls over the sloping vane and back into the channel away from the bank."[1] In the Moultroups' case, one vane gave way entirely; one partly gave way but still did its job, scooping out a pool in the middle of the channel where you would want a pool to be; and the third was still partly intact but did little work because it didn't receive the brunt of the current. Jeff said that he would have to add riprap to try to save his land.

As on many Vermont farms, Jeff and Debbie's pastures formerly extended right to the river banks with little streamside vegetation. In the long run, their most significant conservation efforts may prove to be the new fencing that keeps cattle away from the river, except for one designated crossing, and their intensive grazing practice. The revetment work, funded by a grant from the Lake Champlain Basin Program, was part of a package that included U.S. Fish and Wildlife Service money through its Partners for Wildlife Program to pay for a large part of the purchase of high-tensile electric fencing. The Moultroups had to pay for the installation and agree to maintain it in good repair for ten years.

When I first arrived at the Moultroups' farm, I was greeted by a sign, "Pies for Sale." This was the first hint that Jeff, Debbie, and their children were embarked on a family enterprise that called on all of them to do their part and, as I was to discover, that linked them not only to one another but to their extended family, their neighborhood, and their church. Jeff's father and mother live across the road, and within half a mile up the road he has four brothers, and Debbie has a sister and a brother—Moultroup Valley, Debbie called it. Since graduating from high school in 1976, Jeff cut wood, raised beef cattle, and did odd jobs including backhoe work. Debbie sold eggs and then the pies, thirty-five or forty of them on a weekend. Jeff's father Henry called himself her best customer and Jeff was her "quality control expert," which required a lot of testing. Apple,

apple blueberry, apple peach strawberry, apple rhubarb, strawberry rhubarb, blueberry, blueberry rhubarb, peach, pumpkin, all sorts of custard—"I've tried most of them," Henry said. "I can eat half a pie at a time." Not to be outdone, I said that I could too.

Jeff and Debbie contracted to buy 76 of his father's 350 acres, a former dairy farm, with a plan to return cattle to the land but on a quite different basis. As he spoke of rotational grazing, of the management of pastures, of moveable paddocks, and minimal grain rations, I knew that I was about to see farming in a new light.

Conventional dairy farming for generations has relied heavily on the raising of field crops, especially hay and corn, on the generous application of commercial fertilizers and pesticides, on the feeding of purchased grain, on the use of expensive field and barn equipment, on increasing herd size for economies of scale, and on milking twelve months a year. According to one source, nearly 90 percent of U.S. dairy cows are now confined in sheds, never grazing at all, a practice the grazing literature refers to as confinement farming.[2] It was and is capital intensive, labor intensive, and high maintenance. The major investment in land and equipment, often financed by dangerous debt, has made farmers vulnerable to the fluctuating prices of supplies and milk. The increasing consolidation of the industry in fewer and bigger operations is a familiar story and, to people attached to family farms as a way of life, a sad one. Furthermore, depending on the degree of environmental savvy and concern of the farmer and on the state of government regulations, conventional farms also can be a source of major contamination of our streams from erosion and the runoff of manure and agricultural chemicals.

In Vermont and nationwide, consolidation has gone so far that a major reversal seems almost impossible. The deep trends are familiar. In 1947, for example, there were 11,352 Vermont dairy farms, in 2000 just 1,713. But improved genetics, nutrition, and management have boosted annual milk production from something over 1.2 billion pounds to 2.6 billion.

The nation's tiny farm population does the job with an awkward surplus to boot, such is the effectiveness of conventional, consolidated farming; as a percentage of income the cost of food in the United States is the lowest in the world. It is unlikely that the intensive grazing movement will reverse the general direction of American agriculture, at least not any time soon, but farms like the Moultroups' provide an alternative for farmers looking for methods less capital and labor intensive and more environmentally friendly. There is also a social dimension, an emphasis on family and community life apart from farm labor, which isn't intrinsic to the grazing movement considered as technique alone but which is part of the vision of the movement's prominent spokesmen. I should make clear that I do not intend to suggest that all conventional farmers are large agribusinesses or enemies of the environment, or that intensive grazing need be restricted to family farms. Vermont and other states have many conventional farms that are still family enterprises and many that are taking care to protect the soil and waterways. But the differences between the two methods interest me. Although I did not grow up on a farm, my parents did and their parents before them for many generations.

The dominant dairy cow across the country is the Holstein because it gives the most milk. When Jeff and Debbie decided to go into farming, they chose Jerseys, a smaller breed that gives less milk per cow but more per pound of body weight and is thus more efficient. It also produces more butterfat, commanding a higher price for its milk, and is said to withstand better the summer heat. At first they planned to milk fifty cows, but they now milk fifty-six and will go to sixty. They breed the cows to freshen (give birth) in early April and they milk them until January. The cows are kept outside all the time, summer and winter, except when they are being milked, and in their winter coat they are "pretty fuzzy," Debbie says. In the winter, Jeff provides hay in iron enclosures called hay rings; they can reach the hay, but they can't pull it apart and trample it. The cows can usually start grazing in late April or early May. They receive some corn at the end of March just before freshening but no corn once they start grazing. While they are being milked they are fed 15 pounds of grain per cow daily to supplement the pasture forage. While dry (not being milked) they are fed no grain.

These were radical departures from the practice I was familiar with when I worked on my uncle's farm as a boy. Since grazing is fundamental, pasture management is critical. It is no longer enough just to turn the cows loose to graze all over the lot, hoping they will get a little green stuff. The first thing that must be done is to lay out moveable paddocks with portable electric fencing. Jeff and Debbie's cows usually spend just twelve hours between milkings in a given paddock about an acre in size and then are moved to another paddock. Rotating the cows among some thirty paddocks gives the forage time to grow between grazings. Six to 8 inches is a good height. As Jeff said, "You don't want it too high; they'll just stomp it all down." There is less manure handling since cows deposit most of their manure directly on the land, and pasture does not require the heavy fertilizing of field crops. Although not immune to disease, cows are usually healthier when kept outdoors and are said to produce well through more lactations than confinement cows.

This was all new to me, but it sounded relatively simple until I read some of the intensive grazing literature. I discovered that there are organizations, conferences, books, pamphlets, research studies, periodicals, and web sites offering everything from highly technical advice on pasture maintenance to exhortations to rescue family life from the grueling labor of conventional farming. Some argue that grass-fed meat and milk are healthier for the consumer. The movement embraces both private groups and government agencies. The University of Vermont's Center for Sustainable Agriculture helped to form the Vermont Grass Farmers Association and cooperated with other agencies to establish the Grass Farming Program, which works with farmers and others to promote grass-fed farming. There are Vermont farmer discussion groups, conferences, on-farm workshops, and pasture walks (visiting other farms to observe and learn).

One passionate and deeply knowledgeable expert is Bill Murphy, an agronomist with the University of Vermont's Department of Plant and Soil Science, whose *Greener Pastures on Your Side of the Fence: Better Farming with Voisin Management Intensive Grazing* took me as far into the subject as I was capable of going and then left me behind in its rich profusion of technical detail.[3] It was the French scientist and farmer Andre Voisin, Murphy tells us, who first defined the principles of pasture management for rotational grazing. To reflect the importance of sustained attention to the pasture, Murphy adopts "management intensive grazing" as the most accurate of current terms, for "management is the most important part of graziering [sic]."[4] Voisin's teachings caught on in New Zealand where "farmers feed almost the same number of cattle as there are dairy cows in the United States, and seven times more sheep (plus 1 million each of deer and goats), but they do it on a pasture area the size of Wisconsin! And without grain supplements!"[5]

Although I have had some experience on farms, when I walk through a pasture my eyes and brain simply register "grass." Bill Murphy tells us that in an "undergrazed permanent pasture," one that is just sitting there, there may be ten species of grass or more plus several legumes, sedges, rushes, forbs (broadleafed plants), ferns, and lichens, most of which have potential nutritional value for cattle.[6] That earthworms benefit soil is something most people know, but some of the details are astonishing, such as the fact that pasture soils may hold 1,200,000 worms per acre, while tilled soil would benefit much less, having only 400,000 per acre. When I watch cows browsing, chewing, and resting, I see a pleasant pastoral scene; I do not see the critical relationship among these activities. It is important to have high-quality forage because a cow can't graze more than eight hours without resting regardless of the nutritional value of the feed. Her mouth, 3 inches wide, must take in about 150 pounds of green forage per day—hard work indeed.[7] I certainly never thought of measuring jaw movements and swallowing rates, but research of this kind tells us that short forage, 6 inches high, can be chewed and swallowed much more quickly than tall forage. Since the number of jaw movements per day can vary widely from one cow to another of the same breed (29,150 compared to 18,781 in one study), selective breeding may enable us to produce cows capable of eating more in a given period of time and therefore producing more milk.[8] In addition, the design of lanes and paddocks concerns more than providing enough forage. Lanes should be as short as possible because "milk yield per cow decreases about 2 pounds per mile walked."[9]

These are just a few of the considerations that Bill Murphy discusses and that lie behind the calendar scenes of cows grazing on a hillside. The farmer who is wondering whether to try this method of course asks if it pays. Its advocates say it does—in dollars and cents as well as in its lighter impact on the land. Feed, fertilizer, buildings, equipment, vet bills, and labor costs are said to be lower. One Vermont study found a two-year net income of $579 per cow for grazing versus $451 for confinement farming.[10] Circumstances vary, however, and since actual production per cow may also be lower because of the reduced feeding of corn and grain, the farmer must manage carefully since his advantage lies with reducing costs, not increasing production.[11]

It is obvious, then, that while management intensive grazing simplifies some aspects of farming, it has complexities of its own and some of them are not merely technical. In fact, technical challenges in any field usually are accompanied by a conception of the enterprise that transcends practical success, however necessary such success may be, and this is as true of milking cows along the Huntington River as it is of fishing for trout in the Winooski. To survive, mankind must pluck food from naturally growing plants, kill fish and wild animals, or raise animals and grow crops. There is often a sense of a latent or explicit conflict between the earth and human activities. In *The Metamorphoses* of Ovid we read that in the Age of Gold the earth produced food naturally and was spared the gouging of the plow, whereas in the decline to the Age of Silver men plowed and planted, and "The driven ox whimpered beneath the yoke."[12] In Genesis all was harmonious in the Garden, but after the Fall God cursed the ground and condemned Adam to eat bread in the sweat of his face.

Nature can be seen as either a source of abundance or a dangerous foe as in the ambivalence of the early European settlers of America, some of whom saw a land fertile with trees and game, others a howling wilderness. For Victor Davis Hanson, who believes that the agrarian tradition is the necessary foundation of a sane social order, the job isn't to release some inherent goodness in either nature or man: "No, agriculture—unlike both the squint-eyed exploiter and blinkered environmentalist alike—with divine sanction conquers a sometimes belligerent and reckless nature, improves upon nature, forces for a time nature to serve man, just as the *polis* [the political community] takes the killer, the rapist, the thief, and the arsonist out of us all."[13]

Farmers and promoters of the intensive grazing movement do not all speak with one voice, but they imagine a more peaceful and harmonious engagement with nature and with other humans than the fierce struggle of Hanson's account. The Allen Savory Center for Holistic Management, which appears prominently in the literature, says that in farming or any endeavor we can "work with nature's inherent complexity" (not against it) if we understand the natural processes of an ecosystem.[14] Since "The world functions in wholes," as Savory sees it, "we could never manage a piece of land in isolation from the people who work it or the economy in which both the land and the people are enmeshed."[15] According to Allan Nation, a prominent spokesman and publisher, "Increased profit comes primarily from your knowledge of how to mesh your ruminants with the natural environment."[16] Grazing farms may or may not be organic, but there are significant shared opinions. One organic farming publication quotes Wendell Berry: "An organic farm, properly speaking, is not one that uses certain methods and substances and avoids others; it is a farm whose structure is formed in imitation of the structure of a natural system that has the integrity, the independence, and the benign dependence of an organism."[17]

The assumption is that to thrive one must be in tune with nature. It is a commonplace in nature writing to speak of both competition and balance, the competition under favorable circumstances allowing a relative stability among plant and animal populations. While the grazing movement's environmental sensitivity and feeding methods may be more in tune with natural processes, the very notion of management, here as with fish and wildlife, implies that human beings must exercise control to assure desired results. The word "agriculture" itself suggests the same. Bill Murphy is looking toward an ultimate harmony of agriculture and nature, but to arrive at it, he says, you have to recognize the struggle for existence that never ceases on that lovely, green hillside. Plants compete and gain an "advantage" because of their "selfish qualities"; their "fitness" determines how many descendents they will leave. Unlike Hanson, however, who sees man's interference in this process as a kind of warfare, Murphy speaks a gentler language. To achieve the right mix of pasture forage, the farmer will "influence" the plants' "experience" to transform competition into the "cooperative community aspects" that will benefit cows.[18]

Another aspect of harmony is embedded in the word "sustainable," an idea important to grazing farming but not limited to it; in fact sustainability is a widespread movement encompassing in the minds of its adherents everything from household management to industrial processes. For official purposes, sustainable agriculture was defined by no less august a body than the United States Congress. In addition to providing food and fiber in an economically viable way, sustainable agriculture will protect the environment and natural resources, "integrate, where appropriate, natural biological cycles and controls," and "enhance the quality of life for farmers and society as a whole."[19] The U.S. Department of Agriculture supports a Sustainable Agriculture Research and Education program (SARE) with the Northeast Region's headquarters at the University of Vermont. UVM's Center for Sustainable Agriculture has the promotion of management intensive grazing as one of its "main program priorities."[20] Part of sustainability is of course its literal meaning, for lurking in the minds of many people familiar with agricultural history and practices is the record of those ancient kingdoms that disappeared because erosion destroyed their ability to feed themselves.[21] And here in the United States the Dust Bowl is a living memory.

Management intensive grazing is kind to the soil, but it also purports to be kind to the farmer, to promote a better quality of life as called for by the advocates of sustainability. For one thing, it is said to require less labor. "I enjoy farming much more, but there's really not much left for me to do except watch the cows and the grass to see how they can work together even better," said one farmer.[22] Another said that "With grassland farming most of the work is done in your head, not on a tractor seat in the hot sun."[23] The importance of family and community runs through the literature. Less time working means more time for family and neighbors, and the extra money is welcome too. One advocate said that the test of sustainability is "whether or not we can romance our children into farming" and that to do so farming must be profitable, less laborious, and "fun."[24] From the strenuous, necessary way of life that agriculture was for thousands of years and still is for many in the third world, it has become for us a lifestyle choice, a career opportunity into which young people must be enticed. In the vision of farming to come, the sweat has been wiped from Adam's brow.

It is the tendency of every movement—political, economic, religious, nutritional—to promise not just amelioration but salvation. "Grass feeding," I've read, "is the first step toward the rebuilding of small towns, the restoration of health and the return of that sense of community America has lost."[25] While I had no reason to question the argument that grazing requires less labor than confinement farming and can be more profitable, the assertion that there isn't much to do except watch the cows had the ring of the convert's zeal rather than a sober assessment of the facts. I spoke at length at various times with the Moultroups and joined them for an evening milking. I wanted to see up close what they did, and learn if possible what made them tick as a family, for I sensed from the beginning that they were united by remarkable bonds of love and respect.

I was told that it was rough at the beginning and no easy path since. Before routines were established, Jeff and Debbie sometimes found themselves working until 2:00 or 3:00 in the morning. When the cows started calving, Jeff would go outside a couple of times a night to see if there were any problems. "Jerseys are tender animals," and newborns that seemed weak got special treatment: "We'd bring them in the basement, dry them off with a hair dryer, cover them up with towels. We had a little pen down there we'd put them in. It was interesting." The first year there were a lot of first-calf heifers to break in to the milking routine: "We had to pull some of them into the stalls with ropes."

When I visited for milking, the routines were pretty well in place for both animals and the family. Debbie helped to get the milking started morning and night, but often went back to the house when Jeff finished with other chores such as cutting grass for extra feed or maintaining the equipment. There was less to maintain than on a conventional farm, but what there was needed to work. "The truck was getting desperate," Jeff said. "You had to drag your feet to get it to stop." The four youngsters—Ben, Cherrie, Sam, and Luke, all then still in school—took turns with the milking, two each evening. A flick of the switch before and after every milking turned on the automatic pipeline washer to clean the line that carries milk from the cows to the milk-house bulk tank. But it took vigorous hand-scrubbing to clean the 400-gallon stainless steel bulk tank with a chlorine wash and an acid rinse; the tank holds two days' milk, cooled to 37 or 38 degrees. Lean and muscled with forearms I never had even when I was a farmhand, Jeff's motions were quick and definite. Something of his high school weight lifting still showed or perhaps it was the years of hard physical work; there wasn't an ounce of fat on him.

The milk house was ready. I went with Debbie and Sam to herd the cows into the barnyard where they received some low-protein grain and apples, low-protein because green grass is rich with it. I said that I always thought the Jersey was the prettiest cow. "Sweet face," Debbie said. Sam's job was to work the herd around to get at the grain and then go into the barn to be milked. He urged them forward with the age-old call, "Come, boss," the word "boss" drawn out with a diphthong. (I always thought the word came from the Latin *bos* for cow, but my dictionary says the origin is unknown, dashing my etymological pride.) I noticed a bull in the herd, called Timmy when a calf but now Tim, befitting his happy adult responsibility as clean-up bull, breeding cows that didn't take with artificial insemination: "He's small, but he's doing his job." Cows come in heat every twenty-one days until bred. Cows bred by Tim will freshen late, "But better to have them freshen late than not at all." Jersey cows are usually gentle, but Jersey bulls are known to be temperamental; you have to watch them. So far Tim has behaved himself, as well he should; his predecessor started "growling" at people and was sold, perhaps to end up in a freezer.

Each cow wore a chain with identification tag, but the family knew them all by their markings and by the shapes of their udders, and most had a name. Coriander had a big bag and was a good milker; Misty was "slightly nervous"; and Crystal was a "five-quarter cow," that is, she had a fifth teat that had to be milked along with the normal four. Some of the cows had an attitude, Debbie said, laughing at herself for using a faddish term. Henrietta was the bad girl of the herd. "She's kicked everybody in the family," Sam said, "including Daddy." Jeff said later that training all his heifers at the beginning was no fun: "I don't know about you, but I don't care to be kicked. It doesn't thrill me at all."

Ben brought in eight cows in the order that the cows themselves had assumed, certain animals always pushing ahead, others lagging behind; he and Debbie would each milk four cows at a time. The walk-through wooden stanchions held them with a head gate while Debbie and Ben, having buckled on one-legged milking stools, drew a little milk by hand to stimulate the bag and to check for mastitis, an inflammation of the udder that spoils the milk. They washed the bags, dipped the teats in an iodine-based solution, wiped them dry, then attached the cups with their inserts, called inflations, that sucked out the milk in a mechanical imitation of a nursing calf. The vacuum system made a surging, sucking sound, a steady rhythm. Depending on the cow, the bag was emptied in two to five minutes. You could tell it was empty, Debbie said, when the tubing looked clear; when you no longer heard the swis, swis, swis of the machine; and, Sam said, when the bag got wrinkled as if the cow were growing old. After dipping the teats once more to close the hole to keep out bacteria, Debbie and Ben released the head gate and the cows walked through, down an aisle, and back outdoors to return to the fresh grass of a newly defined paddock.

Debbie did not grow up on a farm, and this was the first year of the dairy. Although short and slender, she had stood up well under demanding physical labor, which had toughened her muscles but had left her feminine, gentle, pretty. It was and is a handsome family, healthy, smiling, energetic. I asked her the questions that in one form or another I asked all of my valley people. How do you see yourself in this life here? What do you think of it all? Do you like the work? "It's hard," she said, "and repetitious, and we've had a lot of frustrations with the drought. Sometimes the well gives out and we have to pump from the river, a lot of extra work. But most of the time it's fun." She made fewer pies now. It's customary to think of farming as a traditional way of life and of entrepreneurial businesses as city ventures somewhat detached from family. For the Moultroups, despite Jeff's rural background, the farm is both a large part of a family way of life and entrepreneurial. It has the hopes and fears of something new. It's an experiment and so far a successful one despite hardships such as water problems, a disease, now under control, that killed several cows, and fluctuating milk prices over which the farmer has no control. In 2001 the farm received the Vermont Dairy Industry Association's Highest Quality Award for its excellent milk.

I asked if the farm was good for family life. Debbie thought it was. It taught the children to be responsible, to work together; it kept them attached to the home place. These things were hard for me to observe directly, but I could see that something in the mix of family and work gave the youngsters a friendly, polite, responsive presence. They were neither shy nor brash. Having struggled to talk with many youngsters whose reticence was almost sullen, it was a pleasure to sit at Sunday dinner with a family whose every member had something to say, could answer questions, and who made me feel at home.

But there was something else, something without which the story would be radically incomplete. One day Debbie said, "God is at the center of our lives, and that's what pulls us through." Early in my travels I had come across "The Confession and Covenant of the Congregational Church of Cabot, Vt." What it declared in 1859, Jeff and Debbie subscribe to today as do their fellow worshippers at the Jericho Congregational Church in Jericho Center, the first Congregational church north of the Winooski River, founded in 1791. The Confession begins "We believe the Scriptures of the Old and New Testaments are the Word of God, and adopt them as a full and perfect revelation of his will, containing all necessary rules for faith and practice; and receive the following as their fundamental doctrines." There follows a list of doctrines, with biblical references, regarding twelve articles of faith including the trinity, the obligation to love God and man as oneself, the creation and the fall, the promise of salvation to true believers, the sacraments of baptism and communion, and the power of each church "to receive members to its fellowship, guard their conduct, and exclude incorrigible offenders." Having thus confessed their faith, the Cabot believers did "solemnly covenant to walk together as becometh the children of God."[26]

When I attended church service with the Moultroups, I found myself carried back a half century to the Congregational church of my childhood just a few miles away in Essex Junction. The hymns had changed and the instruments were piano and guitar instead of the organ, but I heard the old spirit in the new words and tune:

> Shout to the north and the south.
> Sing to the east and the west.
> Jesus is Savior to all,
> Lord of heaven and earth.

Following the offering, there came the familiar doxology but sung by a single male voice, not by the congregation.

Praise God from whom all blessings flow.
Praise him all creatures here below.
Praise him above ye heavenly hosts.
Praise Father, Son, and Holy Ghost.
Amen

Observing people during coffee before the service and listening to them offer individual prayers during the service for family and friends who were sick or in trouble, I thought that it was no small thing to belong heart and soul to a band of the faithful. I sat next to Jeff, my wavering voice carried by his fine, clear baritone as the hymns rose up. During the sermon his strong hand rested on Debbie's.

The Jericho church was then trying to decide whether to leave the various organizations of the United Church of Christ because of the recommendation that local congregations should be "open and affirming" to homosexuals. At a later meeting, 92 percent of the members present voted to withdraw. I tried to dissociate the issue from the fellowship the members obviously enjoyed with one another. I reminded myself that all communities of whatever kind distinguish between insiders and outsiders. With that, I let it go.

Back at the Moutroups' house after the service, grace was said, the dinner served. We talked about the river, the streambank work, the cattle, the drought that had struck again, Ben's first year of college about to begin. Love, friendship, and respect ruled the table with a gentle hand. I couldn't penetrate the mystery of their power, but for a little while I too was under their spell. Driving home I thought about the human need, variously manifested, to acknowledge and celebrate higher realms: wisdom, power, and majesty. We are small; we want there to be bigger things.

17

From Gorge to Gorge

When my feet have gone as far as they can go without slipping on the rounded crest, I lean forward, looking down.
—MY TRAVEL LOG

In the winter, when the trees are bare, you can catch a glimpse of it as you hurtle past on I-89. What you see is an undistinguished hump of rock under the power lines at the Bolton Falls dam of the Green Mountain Power Corporation. What you do not see is the vertiginous drop into the gorge below. Here Ira Allen stood, or so the story goes, the only one of his company brave enough to walk to the very lip and look down, and to this day that modest hump bears the name of Allen's Rock. It is not the most important story told about Ira Allen, but it gave me an idea. Assuming that a place of such attractive danger would be cordoned off by the power company, I imagined a furtive adventure, defying No Trespassing signs and climbing a high fence to put my feet where Allen's were boldly planted over two hundred years ago. Surely some glory would accrue.

So now under the high-tension lines I take a narrow path that leads from the portage trail along a stand of cedars toward the cliff. I see no fence, no sign, and in a few minutes I find myself but a few feet from the edge. Crisscrossed paths tell me that I am not the first to visit since Ira Allen. Below to my left is the top of the dam; I see two men at the powerhouse. Below that, far down and stretching westerly, lies the valley floor, the river broad and silver in the sun, although up close it is brown with silt, for it is running high and a cloud of mist rises above the dam from the plunging water. Farther still to my left, Camel's Hump's distinctive profile rises to the sky. All distances and vistas are grand here, on a vaster scale than the intimate landscapes of much of my river travel. I inch forward. When my feet have gone as far as they can go without slipping on the rounded crest, I lean forward, looking down. These are the Bolton Narrows where the river is squeezed between rock walls to tumble, long ago, over a natural falls and, since 1898, over a dam. Before the dam, the drop to water would have been even greater than now; history deprives me of the opportunity to match Allen's feat point by point. But as I look down, my heart races, I imagine all too vividly what it would be like to fling myself through space, and I decide that whatever the drop, 'tis enough, 'twill serve. I am told that this gorge is the state's second largest, a quarter-mile in length and up to 120 feet high on the south side where I am standing.[1] If I read 200 feet high, I would believe it, and as for standing I am not standing long. Having made my point but still curious, I retreat 10 feet, lay aside my hat, and drop to my hands and knees. I crawl forward and peer over the edge again; it's still a long way down, but my knees no longer tremble.

Only a few times on my journeys have I felt what writers and travelers used to refer to as the sublime. The word has lost much of its force in common parlance. More than simply supreme or excellent, it suggested an awe-inspiring majesty, and unlike the beautiful, to which it was often contrasted, it could evoke feelings of terror as the merely human confronted grandeur and power on a supra-human scale. Although Allen's Rock and the distant landscape below and beyond are hardly the Alps or the Rockies, I am stirred as were the early explorers of those mountains. Another occasion was a late, gray evening in winter, driving east along the valley approaching the high spine of the Green Mountains. Snug and warm in my car, I looked up at the ridges almost black in the dim light and partly obscured by low, clinging clouds. I thought of what it would be like to be up there on the cliffs in the dark, cold and alone. Never had the natural world seemed so utterly other, and the world of light, fire, and human habitation so obviously mine. And yet, such is the peculiarity of the sublime in the imagination, I enjoyed my desolate feeling of alienated insignificance just as I enjoy the swirling currents of fear and vertigo as I look down from Allen's Rock. Or if enjoy is too placid a term, I'll risk a much-abused word and say I thrill to them.

The Winooski, however, seldom lets you forget its human neighbors. Although in plain sight, the dam is momentarily absorbed in the natural scene and I can forget the power lines overhead, but the noise of cars on the highway to the north penetrates my ecstatic revery, and a train whistle and then a glimpse of the train itself, snaking along the south side of the river, fairly shatter it. You could say that Ira Allen, his more colorful brother Ethan, and their cohorts put all this human bustle in motion, and I decide that I will have to pay them a visit before the story is done. But first *The Tub* and I have miles to go.

On my camping journey, the Bolton portage is even more grueling than before. Since I have the wheeled carrier with me, it seems only sensible to spare my shoulders, and a perverse determination to make the carrier work blinds me to the advantages of a more primitive attack. It has rained during the night, but now the sun is out and I let the tent dry while I pack the other gear. After a breakfast of powdered protein mix (less palatable with every meal), powdered milk, granola, peanut butter and crackers (peanut butter never lets me down), and amused by the antics of a beaver who swims around slapping his tail and diving repeatedly, I load the canoe and start up the hill. The trail is steep and slippery with wet leaves. First I discover that I must empty the canoe and carry the luggage ahead in stages with several trips for each stage. But even the empty canoe on wheels is too much for the grade and the footing. I have to belay it with two lines, one tied to a tree uphill to hold it in place while I take a turn around the tree with the second line, tug to gain a few inches, snub tight the slack, and gradually work *The Tub* up to the tree. Down the other side is easier, but again the wheels are of no use because I can't hold the canoe against the steep grade. I take the wheels off and slide the canoe to the bottom. And between bouts with the canoe I stumble back and forth with all my gear. It would be quicker just to carry everything, trip after trip, including the canoe, but I have wheels and I am determined to use them where possible. The Winooski Valley Park District guide warns that this is a difficult portage taking a half-hour at least.[2] I master it in two hours and twenty minutes with no rests.

I repack *The Tub* and put into the river below the dam at 10:10, paddling wide of two fishermen. When I planned this camping trip, I thought that compared to the day trips it would be a leisurely excursion, affording more time for observation and reflection. I had no ambition to be other than who I am, but if I had chosen a model for myself afloat it would have been a Thoreau, not some rapids-running daredevil. As it turns out, I am neither. I am more like a minor league voyageur, packing, unpacking, forever paddling on, my only virtue my stubborn endurance.

This is not to say that I am completely oblivious to my surroundings. I share the river with many ducks, three blue herons, some hundred gulls on a bar, six geese, several turkey vultures, and the hated cormorants whose droppings and nesting have driven other birds off Young Island in Lake Champlain and killed most of the vegetation. A federal permit allows the state to spray eggs with corn oil to close off oxygen from the embryo, a procedure now underway, but does not allow the killing of chicks or adults.

Turkey vultures up close on land are ugly, awkward birds—sometimes I paddle by without disturbing them—and their habits, though useful, are not endearing. Master scavengers, they feed on decaying flesh. Scott Weidensaul in *The Raptor Almanac* tells us that after long debate, observers have decided that turkey vultures locate their food by smell, not sight, discovering along the way that for all their gross habits, they do have a certain delicacy of taste, for the appropriate range of rottenness must be present before they are attracted to it.[3] If carrion is too fresh, they cannot smell it; if it is too putrid, they will not eat it. If it is somewhere in between, it is just right, affording them the satisfaction, I assume, that I find in steak medium rare. Sometimes resorting to piracy, "Turkey vultures have been known to harass great blue heron nestlings, forcing them to regurgitate fish, which the vultures then eat."[4]

If their exertions overheat them, they defecate on their legs to cool off, as do their new-found relatives, the storks. New-found, I say, because in the Linnean system of taxonomy, emphasizing morphology, all New World vultures were thought to belong to the Falconiformes order along with hawks, eagles, and some other birds, as indeed common observation would suggest. DNA classification, however, indicates that they are actually more nearly related to storks.[5] Modern science has a way of undermining what our eyes and common experience tell us: the sun does not go around the earth, species are not fixed, the continents move, solid matter is full of holes, and now the turkey vulture is a stork. We are told that there are other similarities to storks including the absence of a syrinx, "the organ of sound production in birds; the best they [storks and vultures] can do is hiss and grunt."[6] Unappealing on land, in the sky vultures soar on the currents with a grace and ease worthy of the noblest of birds. I see a number of them below and aloft but none more striking than two who perch in a dead tree this sunny morning, facing the breeze. They stand erect with wings extended, tails spreading and closing, individual feathers distinctly visible through my binoculars; like a cat waking up, they stretch as if it feels good. Later I will ask several wildlife people what they were doing and among the answers will be that they were cooling off, warming up, and drying their feathers. I have seen a photograph of three of them in this pose with the comment that "Sunbathing is a common raptor activity, especially among turkey vultures wet with heavy dew."[7] I leave it to the experts to decide.

Canoeists are warned of rapids just before a railroad trestle not far below the dam, but I have no trouble. It is hard to imagine that of the eighty-four Vermonters who died in the 1927 flood, this now gentle stream and its tributaries killed fifty-five, twenty-six of them here in Bolton where I am plodding along, yearning for more adventure. For the rest of my trip this day, I encounter nothing but riffles and calm water. Too calm, in fact, for I could use a brisker current, and to break the monotony I pick out landmarks as short-term goals. One is the village of Jonesville, another the bridge at Richmond where I lost my canoe for a time on my earlier trip, then the four-span railroad trestle in Williston where my father used to fish, followed by the North Williston Road bridge near the railroad crossing where four of my high school friends raced a train and tied (and died), and the trestle between Essex and Williston near the haunted house, and finally the dam and power plant at Essex Junction. These are not only canoeing landmarks; they are the landmarks of memory, and some of them I will revisit before the end of the story.

More obvious on the topo maps than from the canoe, the stretch from Richmond to Essex Junction sees the valley floor widen and some of the state's best farmland appear. I have left one landform region, the Green Mountains, and entered another, the Champlain Lowlands. Some writers refer to the entire 40 miles from the Bolton dam to the lake as the Lower Winooski.

In the 7 miles of impoundment behind Green Mountain Power Corporation's dam at Essex Junction, there has been almost no perceptible current. Only the air moves and that the wrong way, for a gusty wind blows upstream, sometimes turning *The Tub* sideways. I finally come to IBM, a sprawling industrial giant where the Teachout farm was when I was a boy, then round a bend and see the dam across the stretch we called Silver Bowl Lake. I pull up at the take-out point behind the house of my junior high school agriculture and shop teacher. Shouldering my canoe for the portage, I lose control and fall with the canoe neatly covering me. As I struggle out, my first thought is to wonder if anybody has seen me. I walk unsteadily at first and my vision is blurred as if my eyes are focusing independently. Perhaps it has something to do with having paddled pretty much non-stop for nine hours and twenty minutes, hoping to reach my next campsite before dark. I could have walked the distance faster, a commentary on the performance of *The Tub* and perhaps that of the paddler.

This portage, though shorter and not as steep as Bolton's, presents the same problems in miniature; the wheels are a hindrance. I struggle up a small hill, then down a path under the Route 2-A bridge and around the power plant back to the river. I don't have time to linger, but I would like to because I spent many hours here as a boy fishing from the concrete platform at the tailrace of the powerhouse or, more often, across the river on the Williston side from the expanse of water-worn rocks stretching some 750 feet downstream from the dam. Known to some as the Williston Gorge, to me it is the Essex Gorge, for the river was the town boundary and Essex Junction was my home. Long before my time, the site was known as Hubbel's Falls, and over the years five dams were built in this vicinity, the first at the direction of Ira Allen. Various mills and manufacturing businesses were established, turning out more than a dozen products including lumber, bobbins, butter tubs, inkwell caps, and electricity.[8]

The closest I come to capsizing in all my canoeing is in the turbulent discharge below the plant. Despite my frantic paddling, the canoe insists on riding the current right onto a rock. Bow in the air, we teeter there for a moment, and I take a deep breath expecting to be dunked, but instead of tipping, *The Tub* slides off the rock and we bump and bang our way along. It's not much for beauty, speed, or efficiency, but it stays upright. In a few minutes I reach a large island, the downstream end of which rises like the prow of a ship above the river. Here is the perfect campsite, the best of the entire journey, high, dry, open, with a view downstream to the west, and a splendid sunset to boot.There's even a small beach instead of the usual high bank. I land, unpack, set up the tent, and eat some sardines and crackers. It's all I want even though since breakfast I have had only dates, figs, peanuts, juice, and water. I am beginning to fantasize about what I will eat on the drive home—cheeseburgers, fries, and milkshakes loom large—but oddly enough I'm not hungry.

Before turning in, I inspect an unfamiliar structure part way up the bank. It's about 8 feet long, 4 feet wide, and sunk 3 feet in the ground with a bench or roost, a partial tar-papered roof, and some chicken wire that could be unrolled to cover the opening. What could it be? The chicken wire reminds me of chickens, and my imagination leaps to gamecocks, roosters bred to fight and kill. This would be a fine place to hide them between fights. It pleases the restless yearnings of a domesticated soul to think that I am sharing the island with something wild and illicit. Later I will tell my friend Carl who says that it is almost certainly a duck blind. He is almost certainly right.

It was ten after nine when the tent went up. It's dark now. I lie down, close my eyes. As I drift off to sleep, my arms are still paddling and I see a river in front of me, quiet water stretching endlessly ahead.

18

The Bobcat Man

Many hunters know a lot about animals, but Patrick takes it to a whole new level. He eats, sleeps, dreams bobcats. —CHRISTOPHER BERNIER, Fish and Wildlife Technician

If you had a big, open field, a cat wouldn't walk across the middle. A fox or a coyote might, or a deer might, but a cat just wouldn't; it's not their nature. They're a-sneakin' and a-peekin'. That's how they live. —PATRICK SONEIRA, The Bobcat Man

It was still dark the morning I pulled into Patrick Soneira's driveway in the home town of my childhood, Essex Junction. The happy heartache of good things remembered and gone—family, old friends—soon gave way to the enthusiasm of the Bobcat Man, a title he deserves as surely as your medical man deserves the title of Doctor. Setting off in his truck with Gus the bluetick hound in the back (female despite the name), I was introduced in Patrick's intense, excited voice to the pursuit of *Lynx rufus* (or *Felis rufus,* some would argue), to its habits and habitat, and to the efforts of wildlife managers and biologists to learn more about Vermont's bobcat population. The conversation required little effort on my part, for Patrick wanted me to understand.

His father was a hunter, so he learned early. It tells us something about the changing times that in the 1960s he could deposit his .22 rifle, with ammunition, in the principal's office and retrieve it after school to hunt woodchucks along the River Road where, twenty years before him, I had hunted myself. Early on he succumbed to the spell of Vermont's wild feline: "I really enjoy going after the bobcat because I learn a lot. They're a mysterious animal."

But this most elusive of Vermont's major predators was not always held in such high esteem. Because they took game and, although seldom, livestock, "They were considered a varmint, a troublesome animal, something you'd want to get rid of." Many states had a bounty, Massachusetts, I've read, as early as 1727.[1] Vermont was the last state to remove it, in 1971. The bounty was subject to fraud, Patrick said, because Vermont paid for the ears, Maine for the nose, and New Hampshire for the tail, twenty dollars each, and if you had connections in other states, profitable exchanges could be made. It also came to be recognized that bounties were ineffective in most places in controlling populations, but the main reason they were eliminated was that public opinion shifted to considering the bobcat a game animal rather than a pest.

Among those pushing to remove the Vermont bounty were the houndsmen (sportsmen who train dogs to track game), repeating a pattern found in the late nineteenth century when the massive slaughter of birds and animals for meat was reversed in large part, as I've noted, by the pressure of sportsmen on state governments to pass and enforce game laws.[2] Not only was the Vermont bounty removed, a short hunting season was imposed, now just twenty-nine days, January 10 to February 7, and an even shorter trapping season, December 1 to 16; an unarmed houndsman, however, may track bobcats with his dog for training purposes from June 1 to March 15, excluding regular deer season. And to many people now there is something deeply pleasing about the survival of the high-end predators, in wilder regions the wolf and grizzly and mountain lion, in tamer Vermont the bobcat. "There's an element of mystery," Patrick said, repeating the word. "You might not want one in your backyard, but it's nice to know that the earth is still a good place for these critters."

How good the place is for the bobcat depends on the climate, the habitat, and the competition from other predators. Vermont is pretty close to the northern end of the bobcat's range. The cat best adapted to the extreme cold and deep snow of Canada is the lynx, believed extinct in Vermont. It weighs about the same as the bobcat or less and has large, furry paws that can support about twice the weight of bobcat feet in snow. Soft snow more than 6 inches deep, Patrick said, makes heavy going for the bobcat, while a hard crust is ideal. To the south, bobcats can thrive in a wide variety of climates and habitats and were once found in all lower forty-eight states and down to central Mexico. They are now rare or absent in some heavily cultivated Midwestern states and heavily populated Mid-Atlantic states.

Hunting in the Winooski Valley and other regions, Patrick has seen a shift in the bobcat population from the deep woods to the Champlain Lowlands with their milder temperatures, lighter snow, and, most important, more of the mix of woods and fields, the brushy edge habitat favored by deer, rabbits, squirrels, birds, and other small prey. "Bobcats will show up at somebody's bird feeder," Patrick said. "They'll show up on a golf course." For some time as farmland went out of production, the mixed habitat was abundant, but over the decades the forests spread and matured, and the edge and dense understory growth were lost. Bobcats are still found high in the hills where Patrick and I were headed that snowy morning, but in small numbers. Compared to many animals, however, bobcats are never very numerous, for both males and females are territorial with home ranges that vary widely with the quality of habitat, and except for polygamous mating and the female's rearing of the young, they live solitary lives.

Bobcats are both hunters and opportunistic feeders. They stalk their prey as their distant cousins the housecat and the mountain lion do, and like them they also snatch anything that comes within easy reach while they are sneaking and peeking, crouching and waiting, or just basking in the sun. A red squirrel may scurry by, a mouse, a snake; a ruffed grouse may alight unaware of the motionless cat. In their more deliberate hunting, bobcats will sometimes kill deer. It's easy to imagine a bobcat killing a rabbit or squirrel, but white-tailed deer have sharp hooves, they run like the wind, and adults in Vermont typically weigh 100 to 200 pounds. Some studies have reported adult male bobcats weighing as much as 68 pounds (probably not in Vermont) and averaging 28 pounds, the female 15.[3] Patrick once saw one that weighed 40, the heaviest he ever took weighed 31 (a long, thin one measuring 5½ feet, toe to toe), and he said that a yearling would weigh only 13 or 14 and "A big one, a nice one, would be like 25 pounds." Unlike the wolf or coyote, bobcats are solitary hunters; the mate isn't there to help bring down a deer. Although an occasional skeptical voice is raised as to whether bobcats do in fact kill healthy adult deer, the prevailing opinion among the biologists is that they do. Patrick agrees, for "I have found some they've taken and there's hardly a mark, just a little bite on the top of the neck." The deer is probably bedded down, but if there is a chase, it's a short one because with just a slight head start a deer will outrun a bobcat, and it is also contrary to a cat's energy-saving nature to chase any prey more than about 60 feet.[4]

Patrick's account of bobcat habitat was amplified by Kimberly Royar, a Wildlife Biologist for Vermont's Fish and Wildlife Department and an authority on furbearers. Energetic and articulate with a friendly manner and bright smile, Kim is equally at home in a public forum discussing the history of trapping and furbearer populations, in the field advising foresters and landowners on wildlife matters, or, where I first met her, in the lab dissecting the carcasses of bobcats, fishers, and otters. Kim described a long cycle of habitat conditions and bobcat numbers. "When you think about it," she said, "the bobcat was a middle carnivore that was here with wolves and mountain lions and lynx and marten and fisher, and was probably here at pretty low levels historically, especially because we had much harsher winters three hundred years ago than we have now." As the climate warmed and all these competitors were killed off by unregulated harvest and habitat loss, bobcat and fox populations exploded throughout the early 1900s. I remember seeing twelve or fifteen bobcats strung up in Underhill in the 1960s, all shot by one man, Bob Clark, whose success as a bobcat hunter merited notice in his obituary in 1998. In fact, looking down on me right now from the top of my filing cabinet is a pelt from one of Clark's cats, jaws spread and teeth bared, a gift to Eleanor and me from my father, an emblem of the tutelary spirit of this chapter. Because of the bobcats' elusive and solitary habits and their small number, there are no reliable estimates of the state's population, but Patrick sometimes goes an entire season without taking even one, and Kim says that it has declined since the middle of the last century, "Not to the point that we're really worried about it. But we're doing some DNA work with UVM that will give us a general indication of numbers of animals as well as whether the population has been fragmented over time by topography or manmade structures." (Preliminary information suggests that this is not the case.)

As Kim and Patrick said, the mature forest is not the only change the bobcat has had to face; he now has competition that has contributed to the decline in numbers.[5] In my lifetime, the eastern coyote made his way to Vermont, first called a coydog in the probably mistaken belief that it was a cross between domestic dogs and the western coyote. There soon followed the fisher, not a feline as the common name fisher-cat would suggest but an extraordinarily fierce member of the weasel family, reintroduced by the Forest, Parks, and Recreation Department and perhaps also making its way from adjoining states. Except in a few city neighborhoods, there is hardly a dwelling in Vermont from whose windows you cannot glimpse the territory of a coyote family—"There's a zillion coyotes in Vermont," Patrick said—and the prevalence of fishers is well known to the many families whose housecats have gone to nourish this superb hunter and fine furbearer. Based on knowledge of the coyote's home range, biologists estimate 3,000 to 4,800 statewide. To add to the impact of sheer numbers, somewhere in its migration the coyote became larger than it had been in the West, something I was to hear more about later.

As Kim pointed out, the present state of affairs reproduces to some degree the conditions that prevailed prior to white settlement when bobcats were not the upper link of the food chain. After killing a deer, the cat eats what it can, then caches the carcass, that is, covers it with leaves and sticks or snow and returns to it for as long as it lasts. After the disappearance of the wolf and mountain lion, there was food in the pantry for many days, even weeks. Now if a bobcat kills a deer or comes on one that died of other causes, he had better eat fast because within forty-eight hours the coyotes' keen noses will have found it, and in a night or two there will be nothing but bones.

The fisher may have joined the feast as well. Even one fisher, Patrick said, will thoroughly scour a territory looking for food: "He travels, he goes like the Tasmanian devil, he's in a whirlwind. That's how he hunts. You'll cross six or eight tracks and it's just one fisher." Where you find a lot of fishers or coyotes, Patrick said, you won't find a lot of bobcats. It's not so much a matter of actual battle as the cat's temperamental avoidance of conflict, its preference for solitary prowling and feeding. He is a fierce fighter as you might imagine if you have ever seen housecats fight, but he prefers to move off. A single coyote might be able to kill a bobcat, but he would probably decide it wasn't worth the inevitable damage he would suffer; a pair of coyotes could kill it with less risk but would seldom have occasion to do so. Patrick said that while fishers probably kill small bobcats, they are not likely to take on a big one—he has seen evidence of battles—but they usually don't have to fight to get their way because the cat will retreat: "A fisher for its size—12 pounds is a big one—is extremely vicious and territorial and one nasty critter. He's enemy No. 1 to the bobcat. Bobcats don't take a lot of pressure from other animals."

Patrick and I scouted the snowy mountain roads of the Winooski Valley, examining tracks in the beam of a flashlight, hoping to find the roundish tracks of a cat with their three lobes at the back of the pad. Canine tracks, the coyote's, say, are more oval, the pad has two lobes at the back, an X-shaped ridge appears, and you would probably see the nail marks. Cats have retractable claws, so you wouldn't see the marks unless the cat was climbing something slippery. Fishers' claws may show, and there is often a bounding pattern. All this assumes snow or wet ground of the right consistency to register a good imprint; often the impressions are blurred, hard for the amateur to read. From the ledges where they enjoy protection and exposure to the sun, cats prowl down valleys along streams looking for food. Unlike coyotes with their acute sense of smell, they hunt with their eyes; they have to go where the game is or wait for it to come to them. Where brush-lined brooks cross roads, you might find tracks, but all too often new houses have been built near the crossings.

The weather was perfect, cold but not below zero, and with a layer of new snow. "Cat hunting is all weather-related," Patrick explained. "When it's wicked, wicked cold, like twenty below, they're not going to be moving. It's easier for them to go on a diet; they can live on a mouse or a mole and save their energy for staying warm." It says something about bobcat hunting that despite favorable conditions, we didn't find bobcat tracks, not along the roads or high up around the ledges where we later climbed. This made it a typical day. Last year Patrick hunted hard, saw three cats, and took one.

His description of the terrain proved all too accurate: "You think you're going to die from climbing up and then you get to a little plateau." Lean and wiry, he gave no sign of flagging, and my aging legs were challenged to keep up. We proceeded cautiously, our voices low. "Cat hunting kind of gets in your blood," Patrick said. "You either love it or hate it. There's not a lot of action. You have to have sharp eyes. I know it sounds corny, but I always say that to be a good cat hunter you have to become just like the cat. You're using your eyes, you're using your brain, you're sneaking and peeking along." We sneaked and we peeked, and although we never found cat tracks, we saw plenty of other things, Patrick almost always spotting them first. Three deer bounded away, turkeys flew out of a beech tree, porcupine tracks led to trees where the bark was gnawed, striped maples showed moose barking (stripping bark for the minerals, it is thought), a red squirrel scampered up the ledges perhaps to become a bobcat's lunch, and later, descending, we found where a coyote had crossed our trail not many yards behind us, curious and keeping a nose on us. A number of the beeches had bear sign, the distinctive claw marks where bears had climbed for beech nuts, a desirable, high-calorie food as the bears fattened up for winter hibernation, consuming up to 24,000 calories a day. High in two trees we saw so-called bear nests, great clusters of branches and twigs that had been pulled in, broken off, and gathered to bring the nuts within easy reach.

Patrick kept Gus the bluetick hound on a leash. Since dogs hunt with their noses, there is a natural trait for the houndsman to work with, but it took art as well as nature to bring Gus to where she was now, a reliable hunting dog. Dogs love to chase deer and coyotes and they will tackle porcupines, so they have to be trained through hours of fieldwork to resist these natural impulses and confine their passion to the rather weak scent of the bobcat. "They would rather chase something strong that had a lot of stinky odor to it," Patrick said. It helps if you can train them in the company of an older, experienced dog. Blueticks aren't the only breed used for bobcats; some houndsmen use redticks, Walkers, Plotts, black-and-tans, or redbones. Patrick prefers blueticks for their cold noses, that is, for their ability to follow a faint scent.

Patrick explained what would happen if we found a track and he turned Gus loose. She would sing out, the baying that is music to the houndsman's ears, and we would follow the trail. Several things might happen then. Gus might be following an old track so that she would never jump the animal. If she did jump one, he might retreat to some inaccessible cranny in the ledges and might even be asleep by the time Gus reached the bottom. If he didn't make for the ledges, his attempts to throw off the dog might be so complex as to tempt you to endow the cat with rational thought. He might circle three or four times, tracking up an area in a confusing way, then take off in a bee line really fast. Sometimes this confuses the dog, and we might have to help her along by finding the track leading away. Sometimes the cat will run through deer in a deer yard or follow a coyote track footprint by footprint: "If you're putting the pressure on them, they don't have time for stuff like that, but if they have any kind of lead on you, they'll play with you." Not really that afraid of the dog, "They'll walk a fallen log; they may go down a skidoo trail, then hurtle themselves way off to throw you. They may even circle around and sneak down the hunter's track." You can tell if a cat's been chased before. If he hasn't, "He'll run like hell, then stop, perhaps in a brush pile and look back and say, 'Huh? Did I throw 'em'?" This is the cat you're most likely to take because he will repeat these dashes and stops, and eventually you may catch up to him. Oddly enough, it is sometimes the big toms, 30 pounds or so, that are most easily caught because they may be back-country cats that have been seldom or never chased. I had imagined that a cat in flight would eventually take to a tree, but I was told that a tree is a last resort, a retreat if caught out in the open; cats seem to know they would be trapped there. They prefer to hole up on a ledge where they are harder to see and get to. "I've chased a lot of bobcats," Patrick said, "and they would usually rather fight on the ground, get caught by the dogs, fight to the death, before they will climb a tree. Vermont bobcats I'm talking about. I hear that out West and in different places they act differently."

Sometimes Patrick carries a camera, sometimes a pistol, but most often he totes a .30-.30 and 20 gauge over-and-under (the rifle barrel on top of the shotgun). If Gus had a bobcat treed or visible in a hole, Patrick might use the rifle, picking his spot to avoid putting a big hole in the hide, but he would most often use the shotgun, for unless fired very close the No. 4 shot would scatter enough to make a number of small, separated holes that a taxidermist could easily conceal in preparing the pelt for a mount or a rug.

In recent years, nearly all Vermont bobcat pelts that were kept were either mounted or sold for rugs; in 2001, thirty-six of the forty-five cats processed by the Fish and Wildlife Department were taken either by hunters or trappers, but just one pelt sold at the fur auctions and that for only $22. I was to learn more later about the vicissitudes of the fur market, pelts that enter the international fur trade to end up as coats or hats or gloves; Patrick recalled that in the early 1980s bobcats here were selling for $100 and out West for $200 or $250. The best market now, he said, is a taxidermist who might pay $50 to 250, knowing that he could sell a rug or mounted pelt for a profit.

Hunting, like trout fishing, is a blend of the primitive or elemental with the technological, and I suppose it has been so since an ancient technician fire-sharpened a stick or tied a rock to a handle. Patrick trudged through the woods as hunters have done from the beginning; his muscles, eyes, and ears were in play as was his knowledge of habitat and game. The spear had given way to the over-and-under, and when we returned to the truck, on that day itself a hunting instrument, he showed me his telemetry device that picks up a radio signal from Gus's transmitting collar. Before this invention was available, hunters spent many hours looking for dogs that had pursued game beyond earshot or were wandering lost over the hills.

We took a circuitous route home, crossing to the south side of the Winooski, scanning the roadside and the river's edge. Patrick drove slowly, always the observer. We spotted a muskrat on the ice at the edge of open water, then a merganser, worthless for eating, Patrick said: "You throw them away and eat the feathers." We stopped to examine the tracks of two coyotes that were hunting along the side of the road.

Back home, feasting on Betty Soneira's coffee and brownies, I found a bobcat museum in Patrick's small study. He had said more than once that although he does indeed hunt, killing is a small part of his fascination with bobcats. If I had any doubts, the hour I spent there would have erased them, for in addition to the bobcat mounts and other trophies there were photographs, books, files of articles and clippings, and most interesting of all, a series of logs of his every excursion to the woods since 1967. On one page he noted that he had seen four deer, four gray squirrels, two partridges (ruffed grouse), eight hawks, 165 crows, one duck, two red squirrels, four doves, thirty turkeys, and thirty pigeons; and he had put 100 miles on his truck. Knowing the tendency of stories to grow with time, he said, "I've heard old men, they say this and that, and as the years go by the tale gets bigger. This way it's accurate." The record includes every bobcat track he has seen, the date, the size of the track, and the location keyed to a Fish and Wildlife Department map showing wildlife management areas; he passes this information on to the department.

I have a photocopy of the page that recorded his most remarkable bobcat sighting. I've read a description of courtship and mating—"a lengthy sequence of running, chasing, ambushing, and bumping behaviors that precede and follow copulation," much like the amours of domestic cats—but some of the information was derived from tracks in snow and much of the rest, apparently, from the behavior of cats in captivity.[6] One March day in 1994, Patrick entered the woods with Gus at 6:25 A.M. and

> walked up the log Rd about 100 yds or so, just over a raise, out of sight of the Rd., saw—2 bobcats! On the edge of a open area about 40 yards from us in some alders, knew what they were up to right off—*they were mating!* the female was much smaller, very dark rich brown colored, the male was double her size, very light gray, like a bluff-color, its features were very big and blocky—very big head, wide between the ears, large feet. The female was half-sitting—squatting-like while the male sat over her, his back end moving but not holding her with his front legs. The female was looking at us alert but, the male showed no interest except for her, in a few minutes she trotted away with the male walking close behind and to the side of her, at 1st I was going to chase them but, on 2nd thought decided to leave them alone. I've run cats before, and killed cats before but, never lucked-out like this before! this was a-once-a-life-time viewing!

Now and then during my day with Patrick he mentioned the work of Kim Royar, Chris Bernier (a Fish and Wildlife Technician), and the carcass processing at the Fish and Wildlife Department's Roxbury lab. "Gross but very interesting" was his description, interesting enough for him to volunteer there recording data or handling the carcasses. He has the required rabies shot for handling because of his job as a veterinarian's assistant. I am not more than ordinarily squeamish, so I decided that while I wouldn't handle carcasses, I would pay a visit. The bobcat that dies in the woods of old age or starvation—Patrick's biggest cat, for example, was nine and a half years old, his teeth were worn down to nubs, and he probably had at most one more year—disappears quickly down the gullet of a coyote or fisher, but those that fall to hunter, trapper, or automobile have a posthumous existence as useful as it is undignified.

The work at the Roxbury lab enables the Fish and Wildlife Department to monitor the health and stability of the bobcat, fisher, and otter populations, the three furbearer species most vulnerable to habitat degradation and over-harvesting. None of the fourteen species of furbearers that Vermonters may legally trap are endangered, and the department wants to keep it that way. Within ten days of the end of the season, trappers and hunters must submit their bobcats, fishers, and otters to a game warden who tags them with information that enables the department to relate trapper effort, animals killed, and location. Carcasses are then frozen awaiting analysis at the Roxbury lab. Some population changes occur naturally in response to weather, disease cycles, and competition from within the species or from other animals. Fur prices will affect trapping pressure. If necessary, the department can regulate the pressure for the long-term benefit of the species and the trapper. As in all fishery and wildlife management, Kim and her colleagues are always looking for evidence of environmental conditions, for where the harvest is intelligently regulated, the greatest threat to wildlife is loss or deterioration of habitat. There is another conservation reason for the lab work. Since pelts from some endangered species could be passed off as bobcat or otter, these animals also receive a federal CITES tag (the Convention on International Trade in Endangered Species of Wild Flora and Fauna). Pelt and carcass tags can be compared. In complying with CITES protocol, the department reports annually on bobcats and otters to show that the species are what they are represented to be, that they come from Vermont, and that the populations are being monitored. Federal tags stay on the pelts until they are processed or sold.

One cold February morning I drove to Roxbury to meet Kim Royar, Chris Bernier, and a few volunteers who had gathered for one of several sessions held during the winter and early spring. A graduate of the University of Vermont in wildlife management, Kim is chairperson of the Furbearer Project, monitoring furbearer populations around the state with the aim of maintaining sustainable populations. Among her other duties, she represents the Project on the Trap Standards Committee, studying humane trapping methods; she leads a wetland conservation project, testing water control structures as a way of limiting beaver related flood damage without destroying dams; and she gives engaging and informative talks to general audiences. As the late Michael Corbin, then president of the Vermont Trappers Association, told me at a recent fur auction, "She's very, very knowledgeable." A graduate in wildlife management of the University of Maine with wide experience in the field, Chris (a jack of all trades, Kim says, praising his energy) organizes the Roxbury lab sessions and joins the department's biologists in a variety of tasks including data management, extensive field work, and special projects.

I found myself in a very modest facility, a small ranch house with offices on the main level and the lab in the basement. On this, my first visit, the team handled about one hundred animals, fisher and otter. After outlining the procedure for me, Kim went to work. The frozen carcasses had a distinct but bearable smell of dead animal, not rotting, just dead. They were weighed, sexed, and evaluated as to condition, the amount of flesh and fat being the chief indicators. "Look at this female fisher," Kim said, "ribs showing, no fat. I would say she's in poor to fair condition." But since visual impressions are subjective, each animal was put on the scale; this one weighed 3½ pounds but would have weighed 2 or 3 more when alive. Males run much larger than females, sexual dimorphism it's called; one weighed 9 pounds and when alive would have weighed 12 or so. Otters are heavier than fishers, a male weighing in at 14 pounds, and there is no significant difference between male and female weights.

The next step was to remove the jaw. A woman pried open the frozen jaws of a fisher with a screwdriver, severed the lower jaw with a bolt cutter, then sliced the remaining tissue with a utility knife. Someone remarked that the carcasses are easier to cut if they've thawed a little, but they're harder on the nose. Each jaw was then tied up in a square of medical gauze along with a numbered metal disk for identification and put in a pot of hot water to "cook" so that the teeth could be pulled. This, I was to discover, produces a more disagreeable odor than the carcasses themselves, something like a strange food your stomach isn't used to. Chris said that when he gets home, he throws his clothes outdoors.

After lunch, Kim dumped out a batch of cooked jaws and the crew pulled the canine of the otters and the fourth pre-molar and canine of the fishers. Kim put the teeth in envelopes marked with the disk numbers, which were keyed to the information on the tags the game wardens had attached to the carcasses, and all the data were entered on a record sheet. The teeth were then sent to Milltown, Montana, where Matson's Laboratory ages teeth from all over the country. Slicing and staining the teeth, Matson's can then count the *cementum annuli,* annual deposits of a bony substance that take the purple stain. Typically the older fishers and otters are nine or ten years old. Most are younger, but they can live longer. It's rare to find an otter or bobcat as old as twelve to fourteen.

In the 2000/2001 season, trappers and hunters reported 45 bobcats, 264 fishers, and 111 otters; in 2001/2002, the take was 44, 479, and 172. These numbers are not alarming. Although the fisher population is probably as high as it has ever been, with so much good data accumulated the department wants to continue the study. While not really worried, the department is keeping an eye on the otter and bobcat numbers, which have declined since the middle of the last century, probably because of habitat loss and pressure from coyotes and fishers.

At the end of March, I attended a session that included the year's harvest of bobcats. Patrick Soneira was one of the volunteers. There were about forty cats, more of them trapped or hunted than killed by cars; some years it's the reverse. The procedure was similar to that for fisher and otter, but with the added presence of William Kilpatrick, Professor of Zoology at UVM, who was in the second year of a three-year DNA study of fishers and bobcats. Chris cut open a cat to sex it. Unlike fishers and otters, the male bobcat does not have a baculum, a penis bone. Neither do the higher primates, which, someone remarked, explains a certain human problem. Although the testicles are outside, they are often removed with the pelt, but if on opening an animal you don't find a uterine horn and fallopian tubes, you know you have a male. Dr. Kilpatrick then removed a liver sample. As mentioned above, Dr. Kilpatrick's study eventually should tell us more about bobcat numbers and possible genetic differences among regions. We should also learn more about the origins of the fisher population.

Whenever Vermont wildlife biologists and interested laymen get together, the talk usually turns to coyotes, wolves, and mountain lions. This meeting was no exception. Why are eastern coyotes bigger than their western ancestors? There are two hypotheses, Dr. Kilpatrick said, for neither of which is there adequate evidence. One is that on their long journey east, coyotes interbred with wolves, probably the red wolf. The other is that on entering a new, rich habitat with plenty of deer and other food, natural selection produced larger animals. What about the proposal, advocated by some, to reintroduce wolves to Vermont and the rest of the Northeast? Kim said that one problem is that we don't know which wolf originally lived in Vermont. The gray wolf may have lived in the mountains, preying on moose and elk, while the smaller red wolf lived in the valleys feeding on deer. DNA samples of the only four wolf mounts in Vermont indicate the red wolf, but the result is not decisive because we cannot be certain that the mounts were in fact of Vermont animals. Biologists are leery of introducing new species, exotics, because of the potential for disrupting established balances among predator and prey, which in turn could upset other relationships in the environment. Even the reintroduction of a species might disturb things. "You've got a predator here, the coyote," Dr. Kilpatrick said, "that's increasing its body size and beginning to behave in family groups, although it's not a pack hunter yet [as wolves are]. Its behavior is beginning to change—it's sort of intermediate—so you've got a natural adaptation taking place. Should you interfere with that?"

And are there mountain lions in Vermont? Doug Blodgett, Fish and Wildlife biologist, lion specialist, and one of the volunteers that day, said that he receives forty to sixty reports of sightings per year, and there are probably many more that don't get reported: "Almost all of them are misidentifications—bobcat, fisher, coyote." "And housecats," Patrick said. "Oh, yes," Doug added, "a month and a half ago. It sounded really good. I went out. It was a cat all right; it was a housecat." The embarrassed reporter could have taken comfort from the story of the Dartmouth anthropologist, a man who had studied lions in Africa but who reported a "burnt orange tabby" as a mountain lion. The cat, on the near side of a wide, shallow depression that was invisible to the viewer, looked the same size as some deer on the far side.[7]

On the other hand, Doug said, "On extremely rare occasions we get really damn good credible sightings," including two by my boyhood friend Gary Greenwood, one of the best houndsmen in the state. Patrick was with him on one occasion when Gary's dogs followed an animal that, although never seen, left tracks that could only have been made by a lion. Doug discounted stories circulating in Vermont and other New England states that authorities deny the existence of mountain lions because they fear they would scare away tourists and would cause political wrangles over logging and development in the alleged habitat of an endangered species. Yes, there are occasional lions in Vermont, he said, but they are almost certainly escapees or released animals. If we had a viable, breeding population, we would find more sign. Out West he spent time with experts who could easily find obvious evidence. "On Vermont roads we kill everything from mice to moose," Doug said, "and we've never had a lion end up on a bumper." The last one known to be killed in Vermont was shot in 1881 by a man who was hailed as a hero for destroying a dangerous pest. The mount is on display at the museum of the Vermont Historical Society where I saw it as a boy and dreamed of other times and places.

For Kim and Chris their work is more than food on the table. "Not a bad way to make a living," Chris said, "and you feel that you're doing something productive." Kim said, "This is a great job. I can't imagine anything better to do. I think most of us feel fortunate that we ended up where we are."

Patrick spoke of the "mystery" of the bobcat, pointing both to some ungraspable quality in the animal and to the wonder in his own mind and heart. While such experiences might seem lost in the messy practicalities of the dissecting room, I found something similar in Kim and Chris. Indeed I found it in all the natural scientists who graciously allowed me to look over their shoulder as they worked. I have no simple explanation why this should be so. Most of them were probably trained in a scientific context in which mystery and wonder had no place. It was as if they were drawn to their studies and work by something both in the world and their own natures that their science, strictly defined, could not account for. Looking in from the outside, I imagined an expanded science that addressed the observer as well as the things observed.

19

The Haunted House

My father used to tell me that he'd wake up in the middle of the night and hear a buggy. They would get up to look and there was nothing there.

All the sand from here blew over there. That's why my grandfathers both left the area. They couldn't eat any more because the sand was in the house. —HENRY TERRIER

Nowhere in the valley am I a stranger, but this is the heart of the territory; this is home ground. And yet when I think of the features of this part of my journey—everything from the trestle to the haunted house—it is the haunted house that stands for them all, a place of ghosts.

Below the Essex dam the river passes some rich farmland encapsulated by urban growth. Here as upstream, the high river banks often conceal the broader setting and make it surprising to learn that the Winooski watershed has more urban river miles, 47.6, than any other watershed in Vermont.[1]

Unaware of a new canoe exit at the Green Mountain Power dam at the Winooski Gorge, I pull out just upstream of the Lime Kiln Bridge and leave *The Tub* in the scrub trees along the river. A short hike takes me past the sewage treatment plant of Saint Michael's College to the railroad, the most direct route back to North Williston and my car. Trudging along the rail bed, I find myself humming a song my father used to sing:

The rich man rides in a limousine,
The poor man rides in a Ford,
The beggar man walks by the railroad tracks,
But he gets there just the same.

Oh, it ain't a gonna rain no more, no more,
It ain't a gonna rain no more.
How in the heck can I wash my neck,
If it ain't a gonna rain no more.

I skip the verses about the butterfly with wings of gold and the skunk that perfumes the air. I feel sorry for the beggar man because railroad ties are too close together to step on every one and too far apart to step on every other, and level strips next to the tracks come and go. And even in Vermont's winding valleys, there are monotonous stretches that go on forever, rails disappearing in the distance.

But I am soon in Essex Junction and as the name suggests, it is appropriate to arrive by railroad, for after a start as Hubbel's Falls and then Painesville, the area was rechristened to reflect its importance as a railroad junction, ten lines at the peak connecting here and at one time some thirty-five passenger trains a day passing through.[2] Today Amtrak is struggling to support two. Connections were not always easily made, however, as discovered in 1865 by one Edward J. Phelps, later President Cleveland's Minister to England, who, if his poem is to be believed, waited several futile hours to catch a train to Boston only to board the wrong one and, some say, accidentally to return to Burlington where he had started. He relieved his frustration with "The Lay of the Last Traveler," which reads in part as follows.

"And long and late conductors tell
Of trains delayed or late or slow,
Till e'en the very engine's bell
Takes up the cry, 'No go! No go!'
Oh! let me from this hole depart
By any route, so't be a long one"
He cried, with madness in his heart,
And jumped aboard a train—the wrong one.
And as he vanished in the smoke
He shouted with redoubled unction,
"I hope in hell his soul may dwell
Who first invented Essex Junction."

This section of track I have walked many times before, usually in the dark; it was the shortest route home from my high school girl friend's house. I stop to rest on the steps of the Park Street School where I attended first and second grade. One day, having seen the morning's headlines, I was able to announce to my teacher that "Sicily has fallen," a moment of glory for the Allies and me. I walk down the drive to Park Street, look north toward the Lincoln Inn and the Five Corners, south toward the river. Floating above the scene—new buildings up, old buildings down or transformed—I see the phantoms visible only to long memory. I buy a chocolate cone at Rick's NY Pizza and resume my hike. Across Park Street the tracks pass the building that once housed the warehouse of the Eastern States Farmers Exchange where my father was manager. I can see him unloading 100-pound bags from a freight car, and I remember the sweet smell of the grain. A bit farther on, the tracks join the main line to go either east and south or north toward the train station, once a grand dark cavern, where early mornings as a paper boy I folded my copies of *The Burlington Free Press* in the waiting room and heard Red Lyons, the lunch counter man, bellow his daily announcement: "Train going north, St. Albans, Saint Johns, and Montreal. Track one." There was but one track.

I go east. I don't know the exact spot, but before I get to IBM I know I have passed the place where long after I left the village two boys assaulted and tortured two young girls, killing one of them. The tracks run right through the IBM industrial complex. It's like walking through the backyards of a city, and I have the odd feeling that I am trespassing, peeking into people's windows, but inside I see not couches and chairs but piping and all sorts of mysterious machinery. In a few minutes I am in fact trespassing, for I come to the trestle over the Winooski, a link between Essex Junction and Williston, where a sign says No Trespassing. There was no sign when I was a boy and if there had been my friends and I might have ignored it, for here was a fine place to play: a narrow bridge high above the water with no walkway or railing, and a curve in the distance that concealed oncoming trains. There are two small platforms where you can step aside if a train comes, but we were told that if you tried to stand there, the rushing train would create a vacuum and suck you in. More than once we had to scamper off the trestle ahead of the train.

I hesitate a moment. It would be embarrassing to be arrested, but to obey the sign would be a small failure; to defy it would compensate for the lack of a fence to climb at Allen's Rock. I put my ear to the track and hear nothing, feel no vibration. With one last glance around, I trot across.

Beyond the trestle, I hesitate to see if muscle memory will tell me what to do. I cross a fence and walk south into the woods. Nothing is perfectly familiar, but something urges me on. I come to a woods road, and then I find it, the haunted house. It was, I thought, a big house, deserted, windows gone, dilapidated but still standing. We would go in, climb the stairs, conjure up hidden dangers. Now small trees grow in the open cellar hole. The foundation is broken and crumbled. An 8-foot section of ridge with tin roofing lies on the ground, the boards rotted, the tin crumbling in my fingers. Fifty feet away are the remains of a small outbuilding, some tin and foundation bricks. The hood of a 1940s truck rusts among the trees. I pace off the sides of the house—about 20 by 24 feet.

I retrace my steps to the railroad tracks. Somewhere around here was the Redmond farm that bought grain from my father at the Eastern States warehouse. In 1923, river ice tore out the covered bridge at North Williston and dumped it on the farm. When after some delay the selectmen tried to reclaim it, Rob Redmond argued that after thirty days it had become abandoned property and was now his. The court agreed and what was once a bridge became a barn.[3]

I first heard this story from a high school classmate, Edward Whitcomb, and it is the Whitcomb farm on Fay Lane that I am now passing as I approach the crossing. The family actually has two farms, both on the river, this one in North Williston, the original one in Essex Junction. Five generations ago, Lorenzo Whitcomb assembled the Essex farm on rich river-bottom land. The earliest Whitcomb I remember was his grandson Robert, my friend's father, an engineer by college training, an excellent dairy farmer, who in 1950 decided to revive a family interest in Standardbred race horses, building a track and raising and training a number of horses that distinguished themselves not only at the local fair but also at Saratoga. My father sometimes joined him there to risk a few dollars. Bob's uncle Jim Whitcomb raised and trained the mare Dariel that held the world's mile pacing record of two minutes one-quarter second in 1903. Ed also trained as an engineer but like his father turned to farming with his wife Polly and, later, with their sons Lorenzo and Onan and their wives. And a sixth generation is coming along.

When I asked Ed what it was like to grow up and farm next to the river in Essex, he first mentioned family memories of the '27 flood, which floated mattresses off beds on the second floor of the house, drowned the pigs, ruined the stored hay, and moved the big barn 4 feet off its foundation. Next he said that his parents "hammered fear of the river" into the children. It was an enticing but forbidden place to play. There could be no doubt that it was dangerous, for "the proof was right there in the graveyard," Bob's brother John who drowned at sixteen. Over thousands of years of deltaic and floodplain deposits, the river made the farm, but as with the Moultroup place the river seems determined to take it back. Ed and Polly said that it keeps getting wider, twice as wide, Bob said, as it was when he was a boy. Over their forty years of farming, they have seen the banks go, they have lost land. The backstretch of the race track Ed's grandfather made washed away and now the one Bob made was threatened. The proximity of the river also requires farmers to adopt pollution control measures that cost money but do not produce a cash return.

Conservation practices for the Whitcombs are more than a matter of legal requirements, however; they are a matter of principle. They seed drainage ditches to reduce runoff and they are trying to control riverbank erosion. Polly was the first Agricultural Director of the Vermont Land Trust, advising the Trust on what land to try to preserve for agriculture and educating farmers about what it means to separate development rights from their land. She also served as chairman of the Winooski Conservation District, working mainly in the past on water quality issues in agriculture but with a new emphasis on urban and suburban problems. This district includes some 95 percent of the Winooski watershed (and a little beyond) and about 35 percent of Vermont's population.

Conservation is just one aspect of a complex, integrated farm operation that involves all the adults in the two generations and has won praise and honors, including the New England Green Pastures award as the state's Dairy Farm of the Year for 2001. Onan manages the herd and hires employees, and his wife Mary raises the calves. Lorenzo manages the machinery, crops, and feeding. Polly handles the office and bookkeeping and deals with outside agencies. They host between thirty and fifty groups for farm tours each year. They love the way of life, but they never forget that they are in business as is reflected in their corporate organization and its name, the North Williston Cattle Company.

There are striking contrasts to the Moultroups' farm and to management-intensive grazing generally. In addition to the family, the Whitcombs employ three full-time and three part-time workers plus seasonal help. Between Essex Junction and North Williston, they manage about 500 acres of land. They have Holsteins (not Jerseys), some 250 of them, and they milk them year around. Except for two months a year when a cow is dry (not being milked), the cattle never graze; they are fed a carefully calculated diet that includes a lot of grain. The farm uses commercial fertilizers and herbicides, and it grows enough corn to feed its own cattle with a third of the crop left over to sell. It is an admirable example of a family business blending ancestral memories with current practices. Like all farmers, however, the Whitcombs are subject to the vicissitudes of the market, a market that despite government programs is often unkind even to well-established, knowledgeable operators. And who knows what the future holds? I heard recently of a proposed 40,000-cow "dairy park" in the West.

When I set out to explore the valley, I renewed my acquaintance with Ed and Polly, talked about old friends, walked some of the North Williston land, recalled visits to the haunted house, and learned that Ed and Polly had once sat on one of the trestle's little platforms and had not been sucked into the train. Threads of connection ran everywhere. Then one day I received a call from another Essex friend. There had been an accident. Ed was working on a building project at the farm when a beam slipped and struck his head. After eight days in a coma, life support was removed and he died.[4]

Soon I reach North Williston Road where my car is waiting, but as I leave the tracks, December 7, 1953, comes to mind as it always does when I see this crossing. That evening we heard the sirens go by our house on Maple Street. The next morning my father came up to my room to tell me that there had been an accident, train and car, and in the car were four boys from my high school: Paul, Dave, Pat, and George. All had died. I often think of the exuberance of those final seconds as headlights crossed with headlight on the tracks, and the decision was made not to yield the race. For it was a race, or so the story had it.

I have one more No Trespassing sign to defy before the day is done. When I drive back to Saint Michael's, I find that the treatment plant road is posted, but I can't abandon *The Tub* on the river bank. I drive in, then walk and carry, and soon I am back on the public roads, another criminal trespass behind me.

Ed and Polly Whitcomb tell me about eighty-year-old Henry Terrier (who is to die before my story sees print), a neighbor of the Williston farm who lives on Chapman Lane not far from the bridge and the crossing. Henry has a stream of stories about North Williston. Where near the crossing there are now just a few houses and the farm, there was once a thriving settlement, a "bustling railroad village."[5] Here were found a sawmill and grist mill owned in part by one of Ed's ancestors; a train station, a creamery, a post office (four deliveries per day), a telephone exchange; a business that sold hardware, groceries, clothing, farm machinery, and trucks; factories that made cheese, butter tubs, spools and bobbins. Some of these enterprises Henry remembers from childhood, but most were gone before his time. The most famous of all was a cold storage plant built in 1876, said to be one of the first in the country. The plant burned when he was five or six years old, but he heard stories about it, including its ingenious refrigeration method that would freeze meat and poultry using ice and salt: "You've made ice cream, haven't you? Well, it was the same idea." Among the products shipped to Boston, New York, and other cities was so-called western beef. Catching something in Henry's tone, I asked if it was really western beef. "Well, in a way it was," he said. "This was west of Boston." My drive around the neighborhood with Henry confirms what I have read, that except for the former store building, "Every stick and each brick of the business establishments have long since vanished except for a cellar foundation here or there in the tall grass."[6]

On one of my excursions with Chuck Ratté, we visited a sand and gravel pit off Fay Lane near the Whitcomb farm. Pointing to his surficial geology map, Chuck noted that we were standing on glacial sand that probably melted off the end of the ice and was deposited in Lake Vermont as a delta. This is the sand that looms large in Henry Terrier's account of his family's history. Overgrazing led to lower organic levels in the soil, and this in turn meant even less vegetation until eventually the sandy soil was exposed in so-called sandblows, a process accelerated by lower than normal precipitation. (As Henry remembers, it was sheep that caused the damage). The sand blew here and there including near the trestle where my haunted house stood. This house in fact was owned by Henry's father's step-father Philip Phillion and another abandoned house to the north of the rail line was owned by his maternal grandfather Charlie Jeannette. We are parked in a housing development on a hill above the river and trestle, the surroundings all planted to sand-loving pines. "All the sand from here blew over there," Henry says. "That's why my grandfathers both left the area. They couldn't eat any more because the sand was in the house." As a boy and young man Henry used to pick up milk from farmers: "When we came to the top of the hill when the wind was blowing, we used to have to tie a handkerchief around our face. We had to cover milk cans with a good canvas. Down around the curve there was so much sand you could hardly breathe."

In the 1930s, Henry recalls, the Soil Conservation Service stabilized the area with extensive plantings so that now it is heavily wooded. Another family owned the haunted house for a short time after reforestation, but then it was abandoned again. When I say that as boys we spoke of the house as haunted, he says, "The haunted house fits in. My father used to tell me that he'd wake up in the middle of the night and hear a buggy. They would get up to look and there was nothing there."

Henry winters in Florida, but this is his place, Williston and the surrounding towns. (He reminds me in this intense localism of Hap Hayward of Marshfield). He started driving trucks when he was thirteen and the police looked the other way because although you were supposed to be sixteen to be licensed, they knew he had to support his family. He says that he nearly killed himself working so hard. Much later, among other jobs, he delivered grain for my father. He remembers the story of Rob Redmond and the abandoned bridge, but as he tells it Redmond used the lumber to build a house for his hired man who happened to be Henry's uncle. He remembers the story his father told of the Irish immigrant laborers who were swallowed up when a concrete form broke during the building of the Essex dam in 1917, swallowed up and left there as the crew carried on with the job. Or so the story went: "That's the way he told me."

As a boy, if he wanted to go to Burlington, he would cross the trestle to Essex Junction and catch a trolley to the city. Fearing the supposed vacuum, he would lie down on the little platform if a train came by. His father, caught before the platforms were added, lay down between the rails and the train passed over him. When the family moved here in 1930, an average of five hoboes a day would cross the trestle and ask for food: "They mark where they eat. My mother was a great hand to feed them." He was still living on Fay Lane in 1953 when in their last race the boys tied the train: "I worked with the undertaker to help pick them up. I never forgot it. No, I never forgot it."

The fabric of this little world is tightly woven. One thread leads from Henry's memory of the reforestation to my inquiries into stream dynamics and to a historic experiment on the lower Winooski. *The Streamside Sentinel* prints an old photograph of a badly eroded river bend; then one from 1938 showing the bank being sloped; followed by a 1941 shot of the bank covered with grasses, shrubs, and saplings; and finally one from 1993 with dense vegetation including mature trees.[7] The lower Winooski, in fact, saw one of the earliest large-scale bioengineering demonstration projects in the country, perhaps the very first, conducted from 1936 to 1941 by the Soil Conservation Service (predecessor to the U.S. Department of Agriculture's Natural Resource Conservation Service) in cooperation with local landowners and the University of Vermont.[8] Destructive land use practices had depleted upland soils and exposed river banks, then floods and a hurricane did great damage. Project Vermont was launched to experiment with various methods of bank control and reforestation at ninety-two sites, encouraging at the same time agricultural and forestry practices friendlier to the environment. In the mid-1990s, two Americorps members, Heather Short and Jim Ryan, published a study of the results of the experiment in a document of great interest to future projects on other so-called mid-order rivers, its brief title *The Winooski River Watershed Evaluation Project Report.*

The lower Winooski watershed, according to the report, is the region from Jonesville 33 miles downstream to Lake Champlain (a little more than one-third the length of the river), comprising some 71,000 acres, a 5,000-acre floodplain, a river width from 75 to 175 feet and depths up to 10 feet in summer.[9] In my lifetime, crop and pasture use has fallen dramatically while woodland, commercial, industrial, and residential use has risen. There is an abundance of detail I can only sample. Among Project Vermont's riparian planting of trees, shrubs, or their cuttings were approximately 89,000 purple-osier willows, 545,000 other willows, 31,000 red-osier dogwoods, and 59,000 box elders; upland plantings of various species numbered 2,268,000. Sloped banks, riprap, stone-filled cribs, various deflectors, brush matting—these were among the mechanical measures, added to plantings, installed to protect the banks from water and ice erosion and to fill chutes that threatened to cut off meanders.[10]

Results were mixed.[11] Paddling along, I see occasional remains of riprap and other barricades that failed. The Winooski may be a tired river, as one writer said, but like an irascible old man it has spells of destructive energy, destructive because now it must share the valley with human beings. But as I learned more, I came to understand that overall there were more successes than failures. The more I thought about the dynamic character of the river, the more wonderful it all seemed, the more wonderful and the more frustrating. Because it is a dynamic system, our relations with it can never be settled once and for all unless we want it to be nothing more than a completely channelized ditch, the Roman ideal of an aqueduct, and even that might not work, given the peculiarities of the Winooski.

Project Vermont was intended to introduce environmentally sound methods of control, some of which small landowners could maintain or add to without much if any outside help. A basic principle emerged: "it is absolutely necessary to stabilize the toe [lowest portion] to keep a stable bank," a lesson the 1824 House Inn project had not learned.[12] What usually worked best at the toe, if carefully done, was riprap; most banks have held to the present. But here as always the dynamic responsiveness of the river had the last say, for riprapping transfers the erosion elsewhere, although perhaps to a less valuable site.[13] This is not to say that riprapping was a mistake. It is to say that energy has to be expended somewhere. To sum up, about fifty of the ninety-two original sites were still working at the time of Short and Ryan's report, the wooded riparian buffer averaged 50 feet wider than in the 1930s, serious soil erosion dropped from 90 percent of the watershed to less than 1 percent (helped considerably by the abandonment of farms), and various sandblows saw the return of organic soil to the depth of 3 to 12 inches.[14] I am exploring a landscape transformed not once but several times by people.

One of the ghosts of my boyhood is the illusion that there exists for us a natural world free of human influence to which we can repair for spiritual nourishment. We may be able to clarify or even elevate our thoughts and feelings if we take to the woods and streams, but if there are to be woods and streams that we care to take to, we are going to have to think and work to preserve them. And for the most part they will not be preserved in their natural state but in some blended relationship with our presence. We are forced to be the thinking species, investigators of the truth of things, responsible for consequences.

20

Power

Not to praise or blame, but to see. —MY TRAVEL LOG

The river is an unpredictable animal. —HENRY PARKER

Think of the power behind it. It's hard to comprehend the power of water. —TED BAKER

Think of the patterns of settlement. We live here because of the hydropower. You're running the river through a machine. That's what the business is really about. —MATHEW RUBIN

If for no other reason, my summer days fishing at the Essex dam fifty years ago made hydropower a necessary part of my story, but of course no account of people's use of the river could reasonably ignore it. Fishing and power generation in turn called for a brief account of the tensions between the industry and environmentalists. But since I revisited the valley to look through the eyes of others, hoping that my own might see more clearly as a result, I wanted less in this instance to praise or blame than to see the river from the perspectives of those people for whom it is part of an industrial system.

I have on my desk two old hydraulics and hydro engineering texts that reveal incidentally the sanguine expectations once held of the future of hydroelectric power. The one spoke confidently of the multiple uses of river basins with "full development" as the goal, the only anticipated conflict being that between public and private ownership.[1] The other forecast that with the vast potential of America's rivers "the possibilities for the future are almost unlimited."[2] Neither foresaw the rise of the environmental movement.

The current web page for the National Hydropower Association strikes a different note. Proud of its achievements, the industry nevertheless feels under attack. This trade association is eager to tell us that only Canada produces more hydropower than the United States, that between 8 and 12 percent of American electrical generation comes from hydropower, that worldwide the figure is 20 percent, and that fewer than 3 percent of the 75,187 significant American dams produce electricity. Much more hydropower could be produced in the United States by expanding or modernizing existing facilities and by adding generating plants at existing dams. It usually takes eight to ten years to renew a federal license for a hydroelectric project, a costly process and one that delays the addition of new capacity (more than half of the non-federal capacity is up for relicensing between 2003 and 2017). Hydropower is clean, renewable, reliable, and the impoundments behind dams contribute to recreation and fish and wildlife habitat. "Without hydropower," we are told, "the U.S. would have to burn an additional 121 million tons of coal, plus 27 million barrels of oil, and 741 billion cubic feet of natural gas combined," contributing mightily to air pollution. Proponents also point to hydropower's low cost since the fuel is, in a sense, free once the facility has been built.[3]

Although environmentalists do not speak with one voice about the impact of dams and hydro stations, many believe that they are to some extent harmful, even if a necessary component of a region's power mix. They are said to create warm-water reaches of a river, build up excessive sediments above the dam, retard the flow of sediments and nutrients downstream, block or inhibit fish migration, reduce the number and variety of aquatic insects, and destabilize stream banks with the sudden raising and lowering of water levels.[4] (Some, like Waterbury Dam, release water from the bottom of the reservoir with a beneficial cooling effect, but even this facility is faulted for the rapid raising and lowering of the stream.) Environmental conditions imposed through relicensing have brought about considerable improvement since 1980 when Vermont's Agency of Environmental Conservation judged 79 percent of the hydro dams in the state to be harmful to fisheries, but even with these changes many people still consider the dams to be detrimental. Dams are a mixed blessing at best if you value geologic and hydrologic features in their natural state. One study reported that "Of approximately 30 'large' waterfalls and gorges in Vermont, one-half have been dammed. Ten out of 12 waterfalls and gorges on the 5 largest rivers in Vermont have been dammed."[5]

Against these charges the NHA replies that we need "working rivers"; that the new ecosystems are productive in their own way, not unhealthy; that much has been done to ease the passage of migrating fish upstream and down; and that hydro is the most reliable of the renewable energy sources because it can store its energy in reservoirs and impoundments. It also says that the Federal Energy Regulatory Commission often favors various state and federal resource agencies over the industry in relicensing procedures.[6]

To see the river as the power industry sees it took me back to the water cycle as I had thought about it in "Headwaters," but this time with water not as "the blood of the land," as Alice Outwater called it, but as a circuit of energy, forever expended and regained. The ancients were not so far off when they thought of the sun as a god. As we know, it provides the original energy for all life, enabling green plants to make living tissue out of inorganic matter. But it is also the energy of the sun, working with gravity, that powers the hydrologic cycle, evaporating water from the earth into the atmosphere where it condenses and falls again to earth. If it were not for the sun, gravity would pull all water to the lowest level to which it could flow and leave it there; the Winooski would cease to run.

I usually thought of energy in a vague way as the capacity to do work, and that is true. My hydropower people, however, distinguished among different kinds of energy, one producing another. When I paddled the 7-mile impoundment behind the Essex dam, I was floating on an accumulation of potential energy. As the water spilled over the dam, or flowed through the penstocks (big pipes) to the power plant, the potential energy became kinetic (moving). When it turned the runners in the turbines at the bottom of the penstocks, it produced mechanical energy. The revolving turbines, attached to generators, spun rotating magnets past copper coils, producing electrical energy. Without the sun, no flow, no hydroelectric power.

Puzzling over these matters, I realized that I was using such terms as energy, work, and power, in an imprecise and confused way. A student of elementary physics, which I had long since ceased to be, could have straightened me out. If I start with energy as the capacity to do work, then work can be understood as the transfer of energy as, for example, the transfer from the falling water in the penstock to the turbine. A standard measurement of work is the foot-pound, that is, 1 pound lifted vertically 1 foot. Power is the rate at which energy does its work. We commonly measure the rate of mechanical work in terms of horsepower. One horsepower is the amount of power needed to produce 33,000 foot-pounds per minute. The rate of electrical work is measured in watts, but here, although I can say the words, I can't visualize what happens. One watt is equal to the work of a current of 1 ampere passing for one second through a resistance of 1 ohm.

Just as his guns are an integral part of Warner Shedd's love of hunting and his pumps a part of Ron Mercier's dedication to clean water, so too do the facilities of water power capture the imagination of hydro people. When I toured the Essex plant with Green Mountain Power engineer Michael Scarzello and Power Production Worker (operator) Gary Galbraith, I was told that the dam is 494 feet long and 55 feet high. The power plant itself sits low in the river bed, so there are 66 feet of head. The penstocks, 9 feet in diameter, are steel tubes encased in concrete.

Running at peak capacity, the plant's four hydro generators would produce 7.2 megawatts. There are also four diesel-powered generators for stand-by power, each with a capacity of 1 megawatt. (One thousand watts equal 1 kilowatt; 1,000 kilowatts equal 1 megawatt.) Insufficient water and the need to maintain a minimum flow for environmental reasons (there have been regulatory controversies) kept the average throughout that year to about 4.03 megawatts. To my ignorant ear the figure was still impressive (it equaled 5,410 horsepower), but I was told that it would only serve some 4,200 households, whereas the local IBM plant alone required up to 38 megawatts for which it paid about $30 million.

Where did the rest of the power come from? I later learned that on average Green Mountain Power produces about 4 percent of the power it distributes, nearly all of it from its dams (0.5 percent from wind turbines) and buys the rest. The three largest outside suppliers are Hydro-Quebec, Vermont Yankee Nuclear, and the New England grid. In 2001, a dry year, GMP generated only 2.4 percent from hydro. Electrical power consumption, by the way, usually is measured in kilowatt-hours, that is, the amount of energy developed by 1 kilowatt in a period of one hour. Ten 100-watt bulbs burning for one hour would use 1 kilowatt-hour of electrical power.

Vermonters love their state for intimacy, not grandeur. Searching for further comparisons to the Essex hydro peak output of 7.2 megawatts, I found that the United States' champion, Grand Coulee Dam on the Columbia River in Washington, produces 6,500 megawatts. But this is dwarfed by China's controversial Three Gorges dam on the Yangtze, which will turn out 18,000 megawatts when completed, displacing between 1 million and 1.9 million people as the waters rise (the estimates vary). Many environmentalists see the project as a disaster.[7]

I was viewing at the Essex plant a modern form of harnessed power that had its origins at least two thousand years ago, for many centuries the open water wheel and eventually the more efficient enclosed wheel or turbine. Water wheels are in effect hydraulic motors powered either by the weight of falling water or by the pressure of a stream's velocity or both. Over time, a variety of wheels were invented, their usefulness depending on the conditions of the stream and the site. The horizontal wheel lies flat in the water and is turned by the pressure of flowing water against its vanes. Of the various vertical types, the overshot wheel is moved by the weight of water that flows into buckets at the wheel's top. In the breast wheel, water enters the wheel at the level of the axis and moves it by both velocity and weight. In the undershot wheel, water flows against vanes at the bottom of the wheel, moving it entirely by velocity. There are also horizontal and vertical impulse wheels in which water enters from a nozzle. And these are only broad types, a number of variations of which have been devised by restless human ingenuity.

Efficiency is important. The more energy the wheel can capture from the water, the more efficient it is said to be. Consulting my hydraulics text, I found "the well-known maxim [that] 'the water should enter the wheel without shock and leave without velocity.'"[8] The energy that produces eddies and foam in shock when the water hits the wheel and the energy that gives velocity to the water as it leaves are both lost to the wheel's rotation. I was reminded of what I had read about streams, that a steep, rocky passage with all its turbulence and apparent speed actually may run more slowly than the smooth, unimpeded flow of a downstream stretch. As with my foray into the literature of stream dynamics, I soon came upon material that I could not understand, pages of mathematical calculations to determine the efficiency of various wheels.

On the web site of North Carolina's Waterwheel Factory, which still makes wheels although mainly for landscape and ornamental purposes, I learned that water wheels underwent a number of innovations during the rapid industrialization of the nineteenth century. Because in the 1820s "There was scarcely a good hydro site in England or Europe that had not been harnessed" and "Every inch of head was precious, and every gallon of flow was treasured," innovators such as Jean V. Poncelet were heroes. Poncelet redesigned the least productive of wheels, the undershot, to reduce shock almost to zero and to keep water in the wheel until nearly all of its velocity was expended. Efficiency rose from under 30 percent to 65 to 70 percent.[9]

The sawdust blower at Hap Hayward's mill may have been the valley's last direct mechanical application of hydropower for business purposes. What enabled hydroelectric power to take over was the nineteenth century's increasing understanding of electricity and the development of the water turbine. It was only some fifty years from the experiments of Michael Faraday and Joseph Henry with electric generators and motors to the first hydroelectric plant—some say at Niagara Falls in 1879; others say at a site in Wisconsin in 1882. By 1889, hydro is said to have generated some or all of the electricity at some two hundred power companies. (Coal-fueled power plants came into use in the same decade, the coal fire making steam to drive the turbine.)

The turbine is said to have been invented by the Frenchman Benoit Fourneyron in 1834 (I have also seen the date 1827).[10] The wheel's enclosure in a case enables the water to push against all of the vanes at the same time. As in Hap Hayward's small turbine, there are wicket gates that can be adjusted to let more or less water through the runner, the moving part of the wheel. More compact, powerful, and efficient than the open wheel, adaptable to other power sources such as steam, and capable of many more applications, the turbine has undergone many design variations, the contemplation of which pleases the engineering mind.

The more people I met in my exploration of the valley, the more obvious it became that utility and practicality do not adequately explain the interest they take in their jobs or why in the first place they chose one path rather than another. I met first Mathew Rubin, developer and engineer, who with John Warshow formed the Winooski One Partnership, the independent company that built a new, state-of-the-art hydroelectric plant at the Salmon Hole in Winooski and redeveloped a site in East Montpelier. Financially the ventures made sense because of the federal Public Utility Regulatory Policy's Act of 1978, under which utilities must buy power at fixed, long-term rates from qualified independent producers of renewable energy. But what struck me about Mathew Rubin was the scope of his imagination and knowledge. He spoke of the "history and romance" of hydropower as if it were a kind of poetry. Everything about him was sharp, pointed: his slim, spare frame, his intense gaze, his emotional intensity, his crisp, lucid explanations.

These two power stations were for him chapters in the history of water power from the beginning of time to the present and on into the future. "Think of the patterns of settlement," Mathew said. "We live here because of the hydropower." He spoke of the amazing number of dams built on Winooski Valley streams in the early years, and he knew the various types from the early timber crib—frameworks of hemlock logs (resistant to rot if kept under water) filled with stones—to modern concrete structures arched upstream like some beaver dams the better to withstand the pressure. He said that with the advent of electricity the great textile mills on the Merrimack in New Hampshire and Massachusetts were able to leave behind their mechanical hydro systems, which with "their old belts and their noise—thwacka, thwacka, thwacka—were dangerous, dark, Satanic." Mathew's imaginative grasp of our need for power did not stop with the present. Opposed to nuclear power for environmental and safety reasons—"I met my partner chained to a fence at Vermont Yankee," protesting against the nuclear plant—he saw hydropower as an endlessly renewable source of clean energy. "Some day," he asked, "when fossil fuel gives out, where will our grandchildren get power?"

It was not only the historical sweep that engaged him. He thought in terms of regions and systems, the Winooski Valley itself with its power dams, and the network of waterways, generation, and distribution that makes each power plant a part of a larger whole. He reminded me that although small compared to that of many river basins, the Winooski's productive capacity is vast compared to unaided human power. A person rowing vigorously might exert one-tenth of a horsepower, and if prolonged, "One-tenth of a horsepower will destroy a man's body." Mathew estimated the total capacity of the valley as at least 70,000 horsepower and the total actually developed over time was "extraordinary—a lot of energy slaves" doing our bidding. Ask Hap Hayward, he said, how many men it would have taken to do the work of his little dam and turbine. (Mathew's passion for renewable energy, by the way, has now led him to try to develop wind-power sites.)

If Mathew Rubin spoke with missionary zeal, Henry Parker's account of his work was sprinkled with a Yankee's wry utterance. "You sound like a man who has enjoyed his work," I said, to which he replied, "Oh, I've never hated it too bad." He did enjoy it and when we spoke he had been at it for thirty-three years, coming up through the ranks at Green Mountain Power to be superintendent of the hydro and fuel plants, in charge of maintenance of the generating stations and the company's fleet of trucks with sixteen people working for him. From him I gained a sense of some of the challenges of running a plant and keeping it in repair, for the river is an unruly provider that may offer too little energy or too much and that may deliver it clean or choked with ice or debris.

"I don't think we've got a station on the Winooski River," he said, "but what I've rebuilt it, you know, in thirty years." Rebuilding means disassembling the turbine and generator and perhaps having the blades and wicket gates remachined and new seals put in. The agents that over the ages wear away the bed of a stream also wear turbine parts, those "millions of cubic feet of water going through mixed with the fine sands." Henry explained how the river sometimes overwhelms the trash racks at the intake structure, the closely spaced bars at the top of the penstock to keep major debris from entering the turbine. Small trash isn't a problem, for a hydro turbine "will take an awful lot of whacking." When trash accumulates at the racks, it has to be raked away, once upon a time by hand, now usually with mechanical rakes. I have seen branches, logs, boards, bottles, and plastic jugs accumulating at the racks of Winooski One at the Salmon Hole.

In the fall, leaves were a problem for Hap Hayward, and they can plug the racks at the power plants as well, but the most treacherous opponent, Henry said, is ice. One February, a "massive island of ice" smashed the intake structure at the Essex plant. Much more common, however, is frazil or slush ice, needle-like pieces suspended in the water, which stick to the cold steel of the racks, building up quickly. I was to hear later that they can create a powerful vacuum in the penstock, in one instance with another company causing 600 feet of penstock to collapse.

An innovation I knew little about is the so-called rubber dam, which has replaced flashboards (floodgates). Installed on top of a dam, both raise the water level, but flashboards were designed to break off if the water was dangerously high (another type was designed to flop down when supporting braces broke). The rubber dam resembles a long, cylindrical balloon, its inflation computer controlled. I saw Gary Galbraith lower one section with the touch of a button to let a snagged tree float over the edge. The financial advantage is that you can quickly adjust its height to maximize the water available for the turbine. Many controls are now computerized and some dams are monitored remotely with no worker on premises.

Henry said that in the power business appearances can be deceiving. High raging water doesn't necessarily mean more electricity. A plant can only accommodate so much volume; anything above that amount simply goes over the dam and downstream. And if the river is running high, the tailrace (where the water exits the plant) will rise and reduce the head: "Every foot that goes up subtracts 1 foot off the head. The water simply can't fall as far."

I asked Henry what he thought about when he looked at the river on a day off. First, he said, is the river as a source of power and, next, that "When I first became acquainted with the Winooski River it was a sewer for the local communities, and now it's getting to be a fairly respectable body of water." But he never forgets that "The river is an unpredictable animal." In the occasional violent "ice out" during a temporary winter thaw, great ice jams move downstream carving a path "like somebody was cutting it with a chainsaw on both sides." There also can be unexpected floods. Where it borders his land in East Montpelier, the Winooski one night in the early 1980s was quiet at bedtime. By four o'clock in the morning it was as high as in a spring runoff and Henry could hear the water in a nearby brook "tumbling rocks and stones, and I said, 'Boy, I guess we're going to have some water.'" Heavy rain fell in the hills not many miles away, roads were washed out in some of the upriver towns, and the power company had to scramble to open the floodgates.

When I asked Ted Baker what went through his mind when he looked at the river, he said, "Think of the power behind it. It's hard to comprehend the power of water." Eighty-nine when I talked with him and spent the night at his home in Essex Junction, Ted was old enough to have worked on the clean-up of the '27 flood, Vermont's most devastating. He remembered the heavy labor, filling a thousand bags of sand to stem the flood along the railroad on Bolton Flats, and he also remembered how strong he was then. Although weighing just 140 pounds, "I could pick 100 pounds off the ground and put it over my head a couple of times; now I can't do 40." But he asked me to feel his biceps, which I did; they were still firm.

He was watching boxing on television the evening I arrived, and he had emphatic comments on the skill and tactics of the fighters. All evening and the next morning his remarks about the river and life in general were laced with memories of struggles: childhood poverty, years of hard work, fights and near-fights, baseball, wrestling and boxing, and narrow escapes as a volunteer fireman. Power, skill, tactics, alertness, and readiness—these you needed to survive. Repairing tracks after the flood, he lived in a boxcar and worked with a rough gang of men: "What a bunch! Imagine, we'd work ten hours, then box at night." He and my father were friends, fellow firefighters and town team baseball players, and sometimes they wrestled: "I couldn't throw him." Ted is still remembered by many as Cannonball Baker for his blistering if erratic fast ball. Ray Collins, a former Red Sox player, said, "Baker, the only good ball you've got is the knee ball. Keep it there." The next pitch, Ted said, "I hit a guy in the head."

His boyhood was difficult, the family so poor that he was almost given up for adoption. Every scrap of food was used: "If we didn't eat it for breakfast, we had it for dinner. If we didn't eat it for dinner, we had it the next day." When you have the luxury of a choice of careers and can agonize over your dissatisfaction at work, it is hard to imagine what it is like when any job is a blessing. Like Hap Hayward, he took whatever work he could find—plowing for a farmer, picking potatoes, railroad repair—and still had time to raise hell, as he put it. Then came the opportunity to work for the power company, which he did for thirty-five years, mainly as plant operator. Steady pay!

For many years the river was a constant presence, but although he was a skilled employee, he spoke less about the mechanics of electrical generation than about his respect for the force of water and electricity. Sometimes in flood, pigs and calves would float by in the river and at first you would think they were human bodies. He jumped in the river to rescue a man who had been picking dobsons (hellgrammites) for bait and was caught by a sudden release of water. Both nearly drowned. "It was foolish," Ted said, "but what are you going to do?" One damp day, dusting the switches bare-handed in the control room, he felt the hair rise on his head: "Any closer and the electricity might have jumped. Would have burned me right up." You get careless, Ted said, like his friend at a substation whose head brushed against a 2,400-volt wire. He hit the ground, said "I shouldn't have done that," and died. Ted later started an electrical contracting business, which he ran for twenty years while still working for Green Mountain Power, and one day his brother-in-law, who was working with him, "made a mistake" on a pole and died twenty-eight days later: "You can't make mistakes in the electrical business."

For all this, there was nothing grim about Ted Baker. He was still too spirited, too aware of the ironic, often comic, twists of fate and circumstance to be sour or glum. Life was an adventure, indeed still is, even if now the pleasures are gentler, such as his backyard garden—his "little garden of Eden"—with grapes, apples, pears, peaches, berries, vegetables, ten flower beds, and ten birdhouses. He sometimes hired boys to work for him and learned the truth of the adage that "One boy is one boy, two are one and one-half, and three boys are no boys at all." A widower twice, divorced once, a survivor of two heart attacks and four bouts of cancer, he considered himself a lucky man and said that he still had a mission in life, to give something back to a world that had been good to him. He coached Babe Ruth baseball for twenty years and at eighty-nine he had spoken to a group of juvenile offenders, trying to give them "hope and opportunity," telling them that change was possible, and he wanted to spread his message more widely. He wanted to visit every prison in the state, "But if I can reach just one person, that will be enough." A framed letter from one young man reads in part: "I really appreciate you comeing down here to talk. It gave a little inspiration and hope that I didn't have before."

It was through Ted Baker that Gary Galbraith became interested in hydro plant work, and it was through Gary that I came to meet Dale Bacon, a retired GMP operator. Both had stories to tell of the perils of the work. Gary had seen the whole plant shut down in ten minutes because of frazil, and he had felt the building shake from an enormous short circuit. He was there when a shaft broke in a turbine, bouncing around a 1,000-pound bearing pedestal, wobbling the heavy flywheel 6 inches off true, and shaking loose 2½-inch bolts. If the ends of the shaft had flown loose, "the whole turbine could have broken up."

Old operator habits die hard. Asked what he thought about when he looked at the river, Dale said, "color and debris." Mud season could spell trouble for filters, small tubing, and old wooden bearings. "But I enjoy looking at the river more than I used to," he added. "I swear it's getting cleaner."

It was never far from the mind of any of the hydro people that they were surrounded by immense power, hydraulic, mechanical, and electrical. But it was a misstep, not renegade power, that led Dale Bacon to disaster. Before the day of the rubber dam, Dale and his fellow workers had to walk on the top of the dam to replace sections of wooden flashboards. On one occasion the water had only recently been lowered and the damp surface of a new concrete coating was greasy: "I knew it would be slippery, but hey, I'd been out there a hundred times, you know." He had no sooner told his boss that everyone should be careful when down he went. It was a slide, not a fall, but it was a long, steep slide, some 65 feet because of the curve of the dam, and he landed on dry rock. He had time to think as he fell and he remembered saying to himself, "There, you've finally done it." He dislocated all of his fingers trying to dig into the concrete, and when he landed he smashed a foot and broke his leg and back. "The leg corkscrewed," he said, "and was almost trimmed right off." He never passed out and because of the shock felt no pain. He noticed his wallet floating in a pool, "So I put it in my tee-shirt pocket, and I rolled over again and I could see this leg flopping like a fin." He pulled up his pant leg to show me a shocking scar.

All this he told me without the slightest trace of self-pity, in fact cheerfully and chuckling. He had nothing but praise for the rescue workers who hoisted him up the ledge and rushed him to the hospital. I said, rather lamely, "Boy, that must have been something," to which he replied, "Well, it was a day I probably won't forget." He took early retirement on a buy-out program because he was slower than he used to be: "As the younger people came in, I was always hearing 'Now, now, now, we'll take care of that.' Being bunged up a little, they took it easy on me. It was time for me to move aside. But I miss it, of course, had a lot of years."

21

The Abenaki

World view and the relation of man to nature are perhaps the most difficult and problematical aspects of ethnohistorical reconstruction. —WILLIAM C. STURTEVANT

Plants, animals, rocks, and stars are thus seen not as "objects" governed by laws of nature, but as "fellows" with whom the individual or band may have a more or less advantageous relationship. —MURRAY WAX

European explorers and invaders discovered an inhabited land . . . *They did not settle a virgin land. They invaded and displaced a resident population.* —FRANCIS JENNINGS

As a boy in Essex Junction, I knew that arrowheads (stone projectile points) turned up at the Whitcomb farm and that a local banker had an impressive collection, but I was taught that they were just the incidental remains of traveling bands of Indians for whom Vermont was only a wilderness path between southern New England and Canada. The state was said to have had no resident Indians. Furthermore, except for an occasional individual who identified himself as Indian, I would have said that there was no significant Indian presence in the state in my time. Now I am told that Vermont was the "core of the western Abenaki homeland," that the Abenaki or their ancestors occupied a number of sites along the Winooski River and elsewhere, including at least one with projectile points 10,000 years old, and that the western Abenaki with tribal headquarters in Swanton claim continuous cultural and political existence since precontact times.[1]

Although the new version of Indian experience in Vermont is now widely known, its implications seep only slowly into mainstream opinion and my own consciousness. Reverence for my colonial New England ancestors is so deeply ingrained that my heart is reluctant to acknowledge what my head knows to be true—they displaced, often through usurpation and warfare, a people who had occupied the land before them. Even as I write, several towns in southern Vermont including my own are celebrating the 250th anniversary of their original charters, and I have heard little mention of what had to be done to make the region safe for the invaders. Abenaki readers will find an unintended irony in the title of the first chapter of this book.

These matters came to mind as I hiked the short stretch from the Lime Kiln Bridge between Colchester and South Burlington to the Salmon Hole in downtown Winooski, the city that took its name from the river, crowding the stream on the north side across from Burlington. This was deep in known ancient Abenaki territory, with significant archaeological remains, although they lived many other places as well. As I made my way along the edge of the limestone cliffs of the Winooski Gorge, one of the river's more spectacular features, I asked myself what the Abenaki would have seen in their thousands of years before the Europeans: a deeper gorge, the river running free, a mature floodplain forest the remains of which can be seen in the 32-acre Gorge Island. But mainly I was struck by what they would not have seen or heard: the remains of the limestone quarry, the bridge, the railroad tracks, the cars and planes. I was often startled on a quiet stretch of the lower river by the sudden roar of an F-16 from the Air National Guard field in South Burlington, amazed that a machine so small could make a sound so excruciatingly big. In fact, I found no place on the river, from Coits Pond to Lake Champlain, where human sound could not penetrate. The Abenaki lived in a much quieter world.

The Abenaki story is one chapter in the saga of New World Indians, and it is a tragic saga, for however substantial the accomplishments of the United States and Canada, the price that native peoples had to pay was horrendous. There is no historical accounting method by which the occasional success of casinos or oil wells or even the public reconstitution of tribal government and self-help organizations can be made to balance the devastation of European contact. Diseases passed to a native population that lacked the Europeans' acquired immunity to their worst effects. By some estimates, smallpox, influenza, measles, and a host of other illnesses eventually killed up to 90 percent of the native peoples, one of the great demographic disasters of human history. And of course there was war.

But my chief concern is with the Vermont Abenaki, their recent struggles and their ancient world view.[2] Beginning in the 1970s, the Sokoki–St.Francis Band of the Abenaki began to go public, asserting fishing rights in the Lamoille River, arguing for protection of ancestral burial sites, protesting against prejudicial portrayals of the Abenaki in school texts, standing up to local racist harassment, and arguing for state and federal recognition as a tribe. I am grateful to Chief April Rushlow and Abenaki spokesperson Jeanne Brink, as well as a number of writers, for their accounts of these and other matters. The main reason, it is said, for their earlier reclusive existence, often deliberately passing as white, usually as

French-Canadian, was fear of white hostility. The most dramatic episode was Vermont's eugenics program in the 1920s and 1930s under which, it is alleged, a number of Indians as well as other people were sterilized, technically voluntarily but sometimes through coercion. The prevailing view among the Abenaki and most of the Vermont historians, anthropologists, and archaeologists who have commented on the matter is that although the Abenaki moved around to escape persecution and to earn a living, they have lived in Vermont continuously since prehistoric times under a real, if covert, political and cultural organization, and therefore have met the criteria for federal recognition.

They did receive state recognition under Governor Thomas Salmon in 1976 but lost it three months later under Governor Richard Snelling, and have been denied it since, allegedly because of fears of casino gambling and of land claims that would make it difficult to obtain title insurance for real estate transactions. For years they have been seeking federal recognition, a painfully slow process. Asked to comment as part of the procedure, the state of Vermont produced a lengthy document challenging the Abenaki claims of tribal continuity, the alleged impact of the eugenics survey, and even the Abenaki identity of a number of families in the band. To add to the confusion in the public's mind, there were press reports of disputes and factions within the Abenaki community itself. One could add to the story the irony of their name, for as Haviland and Power in *The Original Vermonters* point out, Abenaki means "Dawn Land People," a name given them by a tribe to the northwest; almost no one calls them by their own name for themselves, Alnobak, "which means 'ordinary people,' as opposed to monsters, giants, and all other peoples."[3]

Curious about the number of Abenaki in Vermont at the present time, I asked Chief Rushlow, who said that she doesn't give out exact numbers but "in the thousands." Haviland and Power said "at least 2,000" as of 1994. They estimated the number of Abenakis in the Champlain Valley before contact at 4,000; another writer has given a precontact figure for the entire state of 10,000.[4] Haviland and Power noted that the U.S. Census of 1970 found only 229 Native Americans in the state whereas in 1975 the Boston Indian Council's Manpower census reported 1,700, the sudden jump attributed to the "resurgence of pride in their ethnic identity" and the willingness to come forward.

As I listened and read, it became apparent that the self-understanding of at least some Abenakis today is traceable in part to an ancient view of the all-encompassing environment we call nature. Nature is a rich word with an intimidating range of meanings. To suggest only four, it can refer to God's creation, which can be read for instructive messages about the divine plan and which may be subject to divine intervention. It can indicate a structured hierarchy of inanimate material things and living beings within which humans can find their proper place and function. It can suggest a strictly physical universe without moral or spiritual implications, operating according to unchanging natural "laws." And perhaps most common in the popular mind, it can refer to all that is not human—rocks, plants, animals—imagined to have its truest existence when untouched by human beings.

With one's eyes adjusted to see the world in a certain way, it is difficult to imagine another view and probably impossible to enter fully into a radically different understanding of things.[5] Although I know only a little of their specialized knowledge, I share with Chuck Ratté the geologist and Kim Royar the furbearer biologist much in the way of general outlook. Although not an engineer, I understand the ambition of the developers and something of the technological principles of Winooski One's new hydroelectric plant at the end of my hike from the Winooski Gorge to the Salmon Hole. And I can easily imagine the Allen brothers' vision of flourishing farms and villages (and handsome profits) as they bought and sold rich land in the lower Winooski Valley. But down through the ages the Abenaki along the Winooski and elsewhere understood their surroundings in a deeply different way from any Euroamerican. (Since much of the evidence comes from closely related tribes and assumes that with care one can extrapolate from historical materials back to the precontact period, I have to rely on the scholars and storytellers who think that various northeast peoples held similar beliefs and that traditional stories and practices provide access, if imperfectly, to a more distant past.)

To put it most bluntly, for the Indians there was neither wilderness nor nature. Of course they knew that swollen rivers or freezing weather could be dangerous; some of their stories tell of threatening mythical (to us) creatures in swamps, lakes, and forest. Obviously they had an intimate knowledge of animal habitat and behavior, of the nutritional and medicinal uses of a wide variety of plants, and of the ways in which tools, weapons, clothing, shelter, transport, and ornament could be fashioned from the materials around them. From an anthropologist's point of view and certainly in the mind of the Abenaki, if such a Darwinian view could have arisen, the Indians and their ancestors had adapted successfully to their changing environment.

As Haviland and Power tell the story, the so-called Paleoindians, Vermont's first occupants, came to a cold country not long released from glacial ice, probably in the tenth millenium B.C.; hunted large game such as caribou, elk, moose, mastodon, mammoths, and marine mammals; and also fished and foraged for food. Over the centuries the climate warmed, Champlain became a freshwater lake, and many species of large mammals disappeared, due in part according to a debated theory, to over-hunting. By about 7000 B.C., the way of life we call the Archaic was emerging, with increased reliance on fishing, food gathering, and the hunting of deer, bear, and smaller animals. According to some scholars, the western Abenaki way of life of early historic times was largely set by the late Archaic period in the second millennium B.C., and persisted through the following Woodland period with the development of pottery and the eventual introduction of the birch bark canoe and the bow and arrow to supplement the dugout canoe and the spear. This was a long period of relative social stability, a way of life that by 1600 found the western Abenaki with "not only more leisure time but also a better diet, a slightly longer life expectancy, and less violence in their lives than did most people then living in Europe."[6]

Given the Indians' successful adaptation to their environment, why do I say that for them there was neither wilderness nor nature? Consider what anthropologists call the subsistence cycle of the western Abenaki. As described by Haviland and Power, the Abenaki lived along major streams including the Winooski in villages estimated to have about a thousand people grouped in family bands consisting of related nuclear families each with its own space in a rectangular bark-covered house.[7] In the spring, the men fished and hunted wild fowl while the women tapped maple trees and collected plant food. In the summer the women continued gathering and tended the gardens while men fished and hunted small game. In the fall, all the able-bodied men and women left the village for hunting territories marked out for the exclusive use of each band. The band in turn hunted only a quarter of its territory at a time to allow the moose, deer, and smaller game to regenerate. In the depth of winter, they returned to the village and ate smoked and dried meat, berries, nuts, and other plant foods, including corn, squash, and beans once limited agriculture was introduced (a more significant source of food in southern New England, decreasing as one went north). In late winter, they returned to their hunting territories. Had they stayed in the village year around and not ventured far afield, the notion of settlement versus wilderness might have arisen. Instead of venturing into a "wilderness" as the early Europeans thought they were doing or even as a modern hunter may feel if he hunts the back country, the Abenaki were taking food in familiar territory. As one scholar has said, the Abenaki "may be studied in the library and in their homes, but they may be understood only in the forest."[8] (The Indians' extensive burning to clear fields and create brush habitat for game probably was not practiced in the interior of northern New England.) It seems unlikely that the Abenaki would have associated the woodland with the cluster of meanings the word "wilderness" brings to my mind—uncharted, mysterious, fundamentally non-human, a place of wild creatures independent of humankind.[9]

There is a more profound reason, however, why the European notion of wilderness would have been foreign to the Indian, a reason that takes us to the question of "nature." Although there were apparently many variations over time and among different groups of Indians, one probably can generalize and say that to the Indian not only were animals and plants alive in the most obvious way, but at least some of them had a spiritual existence as well and belonged to a universe of beings many of whom were related to one another in ways not imagined by the Europeans. Even rocks and mountains and lakes were in some sense alive. This is what the anthropologists call animism, the belief that spirits or souls are found in all things. As one writer says, the northeast Indians didn't live in a natural world of plants and animals and the physical phenomena we study in our various sciences but rather in a social world of related living beings with themselves at the center, responsible for maintaining it through ritual.[10]

Hunting, in this view of things, was more fundamentally a spiritual than a physical undertaking;[11] it was a "holy occupation."[12] Animals were thought to be intelligent and to yield themselves willingly to the hunter if proper rituals were observed in preparation for the hunt, in its conduct, and in the disposal of animal remains. Preparation might include inducing dreams through fasting, drumming, singing, and sweat-baths, which could then be interpreted to predict success in the hunt. Sweating also purified the hunter to enter the spirit world of hunting. Some peoples practiced divination through the interpretation of animal bones that had been burned to produce marks indicative of, among other things, various animals, locations, and success or failure in pursuit (scapulimancy, a practice apparently similar to the various kinds of divination found among the ancient Greeks and Romans and many other peoples).

The actual killing of the game was a subordinate matter, assumed to follow if the correct spiritual elements were in place, but proper conduct was required in that respect had to be shown to the animals, forgiveness might be asked, and killing for sport or beyond need might be forbidden. And after the various parts of the carcass had been used for food or other purposes, the remains had to be disposed of respectfully in a way that would please the spirit of the animal. A beaver's bones, for example, would be returned to the water. If any of these or many other rules were violated, animals would not let themselves be killed or their spirits would not be reincarnated in new bodies for the further sustenance of the Indians. If both animal and Indian did their part, they would enjoy "a relationship of mutual courtesy: intelligent animal beings and intelligent human beings had contracted long ago not to abuse one another."[13]

Both animals and humans were thought to have both a body and a spirit or, as Haviland and Power call it, a vital self. The spirit could leave the body and return as it was thought to do in dreams, which were believed to be actual events. Some people, usually men, had exceptionally powerful and active spirits, which, under certain circumstances, could predict or control the behavior of animals and other humans; they were the most adept at divination. These were the shamans or medicine men.[14] They could heal, but they could also harm, and they were not always benevolent. While it is customary among modern writers, speaking out of our scientific knowledge, to refer to the role of the shaman and the general understanding of which it was a part as belief in the supernatural, to the Indian the distinction between natural and supernatural would have made no sense. The body was no more "natural" than the spirit. (Some writers note a certain irony in the scorn for Indian superstition on the part of those Europeans who believed, among other things, in the virgin birth and the creation of the world in six days.)

As to the broader concepts of the world, how it came into being, we are told that for the western Abenaki there was the Owner, Tabaldak, creator of all living beings with one exception. Among his creations were harmless tiny creatures who lived underwater, a dangerous swamp spirit who could tempt children into the swamp, and two much more threatening underwater beings, a big lizard and a big snake. These creatures turn up in Abenaki stories told and retold down through the years and tell us that although the Indians did not have a wilderness in our sense of the term, their world had its dangers.

The one creature the Owner did not create was Odzihozo, "He Makes Himself from Something," a powerful, self-created being. Odzihozo is one of the so-called transformers found in the stories of all the Algonquians, the family of peoples to whom the Abenaki belong. He is called a transformer because although he did not create them, he had the power radically to change the earth and its creatures. As one scholar tells us, no matter the particular name or characteristics, "a transformer is that personage who was instrumental in shaping the face of the earth, modifying the elements, and altering the biological species, usually from giant protospecies to their present size and form." It was Odzihozo who heaped up the mountains, dug the Winooski and other rivers, and scooped out Lake Champlain. His final act was to transform himself into a stone atop a rock in Champlain, known to most Vermonters as Rock Dunder, where he could view his beloved lake for all time: "He still likes it, because he is still there and used to be given offerings of tobacco as long as Abenakis went this way by canoe, a practice which continued until about 1940."[15]

But Chief April Rushlow told me that even today, some sixty years later, Odzihozo has not been forgotten by his people. April Rushlow is the hereditary chief of the Sokoki–St. Francis Band of Abenaki. She presides over the band with a commanding voice, strong opinions, and fierce devotion to the Abenaki cause, traits that she no doubt acquired from her father, Homer St. Francis, the outspoken, controversial, long-time chief before her. In her office at the Abenaki Tribal Headquarters in Swanton, she graciously answered my questions and told me what she could of practices and beliefs today. I knew in advance that certain beliefs and ceremonies are confined to band members, and I didn't press, but when I asked about Odzihozo, she said, "You'd be surprised if you went out there how much tobacco is there. A bundle like this," and she held up a small prayer packet to show me.

I knew how important the lake, the Winooski, and other rivers were to the Abenaki for food and transportation. "All water for native people is considered holy water," Chief Rushlow said. "If we don't have water, we will not live, so we try to protect all areas with water. In our history and still today, Mother Earth is respected in every sense of the word. Anything we take from Mother Earth we use all of it whether it be a fish, an animal, whatever." If there are unusable remains, they are either buried or left for other animals to eat, and tobacco is put near it as a sign of respect.

Like many of their fellow Vermonters, the modern Abenaki care for their surroundings for reasons of health and environmental protection. They sponsored a Riverkeepers project under government grants, for example, to test water for pollution, and since they try to include traditional foods as part of their diet, they worry about elevated mercury levels in fish. Blended with this contemporary understanding of the natural world, however, are elements that have come down from ancient times, beliefs regarding the spiritual efficacy of certain plants, the existence of animal spirits and totems, and the importance of proper burial.

When I asked why tobacco is used as a sign of respect, I was told that it is one of four sacred herbs along with sweet grass, cedar, and sage. "Sweet grass is used in our spiritual belief system," she said. "There are certain times that we use it or burn it. I won't get into that though."

Speaking of hunting and fishing, Chief Rushlow said, "I was born and raised to do that, and gardening and gathering." She elaborated on the respect that must be shown to the remains of fish and animals, pointing out the deep difference between hunting for sport, not the Abenaki way, and hunting for food. Respectful treatment is not simply an acknowledgment of the usefulness of the fish or animal, nor is it only an ecological matter of contributing to the food chain. Behind it is the belief apparently inherited from the ancestors that "The spirit of the animal is kind of like our spirit. Everybody has a spirit and in order to release that spirit, we do prayer and put tobacco." Abenaki families have animal totems, I was told, which are passed down from generation to generation. I have read that the animal may have been an especially important food or may signify a family trait such as wisdom (often associated with the bear).[16] Chief Rushlow said that the turtle is important to all Native Americans because there are twenty-eight marks on the outside of the shell, representing the twenty-eight days of the Indian month, and thirteen on the inside, representing the thirteen months of the Indian year. The moon determines when certain things such as planting and gathering should be done.

When we spoke of burial, I remembered the love and sorrow with which my parents were laid to rest. But there is another dimension to the burial customs of the Abenaki who believe that the community isn't healthy if people are left unburied or are later exposed by, say, contractors or archaeologists. The spirits of exposed ancestors can make their descendants ill, Chief Rushlow said, adding with a rueful laugh that she didn't know why it should be the descendants who become sick and not the people who disturbed the graves. She herself has suffered. Some remains have been recovered from museums and reburied. Contentious issues have arisen over burials violated during construction on private land, and the state has bought several plots of land to preserve them as burial sites. Chief Rushlow is indignant when people say, "They are just a bunch of old bones. I ask if they would want me to dig up their grandmother." She often goes to burial sites to pray and put down tobacco.

From early times, those native people who resisted total absorption into Euroamerican culture have lived in two worlds. European trade goods including cloth and metal implements were quickly adopted, but used first for traditional purposes. The merging of Indian beliefs and practices with those of the French missionaries was common. Eventually Indians had somehow to blend ancient and modern views and ambitions. If they could not, they would either lose their native identity or they would have to withdraw from a contemporary life much of whose worldview they had already absorbed and whose benefits they wanted. This is evident in the revealing work of Frederick Wiseman, Abenaki scholar, writer, and Director of the Abenaki Tribal Museum and Cultural Center, whose book *The Voice of the Dawn: An Autohistory of the Abenaki Nation* is recognized as the first scholarly book about Abenaki life written by an insider. When I asked Chief Rushlow about religion, she said she considers herself Catholic even though she was never confirmed and was married in a native ceremony: "To me the native way and the Catholic way are similar; they can blend. There's only one God, I don't care what community, only one God, one Creator."

The attempt to blend reaches beyond church to nearly all aspects of life. In its self-help efforts, the Sokoki–St. Francis Band struggles for a cleaner environment as measured by modern science, it wants more and better education for its children, it hopes for better health care, and it sees the need for more economic opportunity. At the same time, in its own view, its identity as a people depends in part on passing down an ancestral heritage based on quite different assumptions about the world. It wants its children to remain Abenaki even as they master the skills of contemporary American life.

In all my travels in the valley, my pleasant task has been to try to see the region through eyes not my own. Nowhere have I found a vision more alien than in my imaginary journey to the world of the precontact Indian. Abenaki hunting customs apparently made them practical conservationists, for by allocating and alternating hunting territories they allowed for the regeneration of game. But many native peoples either never followed this practice or radically deviated from it under the pressure and temptation of trading furs for European goods, for the decimation of animal species, particularly in Canada, was carried out mainly by native trappers and hunters.[17] And the understanding of animal-human relations among the Abenaki was sharply different from that of modern conservationists. Modern wildlife management makes assumptions about animals utterly foreign to the Abenaki understanding. Animals in this view are in effect a crop to be harvested just as corn and oats are; "harvest" is in fact the word that managers use. The very notion of managing is remote from the spiritual relationship that was believed to exist between animal and Indian. The spiritual unity of animal and human life in the Indian view is inverted in the material unity assumed by modern scientific thought, important strands of which hold that human beings are only highly complex animals.

The origin of precontact Indian beliefs is lost in antiquity. Anthropologists tell us that people have to make sense of their world, both their physical environment and the fact of their own existence, and in the absence of science they will devise explanations that to them are coherent and satisfying. The test from this point of view is not what we would consider the truth but the social and psychological effectiveness; the belief system is seen as an adaptation not so very different from the adaptations people make to the physical environment. People like the Abenaki who lived mainly by hunting, fishing, and foraging, tended to see themselves as an integral part of their environment as is reflected in their beliefs. This makes their way of life appealing, at least in the abstract, to some modern environmentalists who see people's detachment from nature as unhealthy for themselves and dangerous to the environment. The prevailing modern view, on the other hand, is that with the aid of science we can achieve an ever increasing mastery of nature in the interest of health and comfort.

The anthropologists may well be right that it is the human need to understand our surroundings and our own being that gives rise to such belief systems as the Abenaki's. As an observer, however, I am happily free of any anthropological or scientific responsibility. I can speculate about another dimension. I imagine what it may have been like at the dawn of human awareness, watching trees reach toward the sun, seeing fish swim, birds fly. I've noticed that whatever their worldview or scientific training, the fishermen, hunters, trackers, and biologists of my acquaintance never quite respond to fish and animals as mere complex mechanisms. If we are honest with ourselves, mustn't we admit that somehow life calls to life, that we feel a kinship? What effort it must have taken over the ages to reduce that kinship to biochemistry and electric circuitry.

My mind swings between two world views as I make my way through this short stretch of the ancient Indian heartland, my heavy step and the impress of modern life bearing down on territory once marked with a lighter tread. It would be foolish to pretend that I do not appreciate contemporary comforts: gourmet breakfasts at Sneakers, hearty pasta dishes at Papa Frank's, a welcome portage in the van of Mark Snyder of Priority Express who, on my camping trip, carried me and *The Tub* through city streets to the canoe access below the Salmon Hole. And it would be less than candid to pretend that the sufferings of the Abenaki prevent me from marveling at this microcosm of Indian, Euroamerican, and natural history.

I have found no other reach of the Winooski that so vividly epitomizes its mixed character as a natural phenomenon and a human resource as the mile or so that terminates in the splendor of the Salmon Hole. It is here that the Abenaki gathered for spring fishing, that Ira Allen set in motion the forces of industry, that the early woodsmen brought the logs of the virgin forests for shipment, that cloth was manufactured for 150 years. Here entrepreneurial zeal often was frustrated by the river's natural ebullience, most strikingly by the '27 Flood that swept away the highway bridge and damaged the mills to the tune of a million dollars. Here industrial chemicals were added to sewage, befouling the river, and here the waters have been (partly) restored, cleaner if not pristine. Dams will never be admired by lovers of natural streams, but here we have hydropower with a conscience, the Winooski One plant sunk out of sight in the ledge, its fish lift part of a trucking operation to move steelhead trout and landlocked Atlantic salmon upstream to spawning waters. Failed textile mills have been converted to business and residential use. There is no riparian buffer where the mill buildings are washed by the river, but Gorge Island, Winooski Falls Park, and the Winooski Valley Park District areas upstream and down give people access to natural magnificence in an urban setting. Most, perhaps all, of the species of fish that lived here before the Europeans came continue to exist, if under more difficult conditions. There are wilder stretches of the river and a few of equal or greater confinement, but nowhere else for so long a time have the human and the natural met in such an intense engagement of destruction and cooperation.

22

A City of Mills

The Superintendent and the Doctor

The story of Winooski begins with a river.

—VINCENT FEENEY, *The Great Falls on Onion River: A History of Winooski, Vermont*

If I was to start my work life over again, I'd do it pretty much the way that I've done it. I enjoyed textiles. There is an awful lot of mechanical work, a lot of ingenuity.

—RAYMOND ROY

My mother would say, "I'll do everything in my power so you won't work in the mill." But nine people were interviewed for a video about the mill and there was only one negative comment.

—RITA MARTEL

That's the reason [a good childhood] I came back to Winooski to practice. I felt I owed the place something.

—DR. ROBERT E. O'BRIEN

For Raymond Roy, a mill superintendent, the workings and life of the Winooski mills held the same fascination as the complex dynamics of streams do for Lori Barg or the intricate habitat requirements of the bobcat for Kim Royar. It was Raymond Roy who helped me to see a textile factory as an integrated system but even more to see how deeply satisfying such a system can be to a certain kind of man. For him, each ingenious process in cloth making had its special interest, but all came together in a whole that was intriguing and beautiful. Of course, all experience is partial. Raymond did not know at first hand the dexterity, fine touch, and strain of running looms in the din of the weaving room. And he had nothing directly to do with national and international markets and finance, although he was well aware of those perilous seas that floated or sank factories because of forces far removed from the local scene. But his vantage point was just right for a view of the mill as a complex enterprise, vital to the life of the city, where a man of spirit and ingenuity could make his way in the world.

I gained a wider view of the community through the eyes of Dr. Robert O'Brien, son of the mill manager and Winooski physician for half a century. Cousin to Raymond Roy—their mothers were sisters—Dr. O'Brien's parentage, childhood, education, and profession wove together the threads of Winooski's life in a pattern as compact as the designs produced by the mechanical marvels of a loom. Through Dr. O'Brien's eyes, I saw a community that had existed side by side with mine for over two hundred years, the two appearing similar from a certain distance, but up close each seen to have a composition and currents of its own. Talking with him I also heard a benevolent irony about human affairs and sharp opinions about medicine, politics, and the frailties of the human heart, all gentled by the true physician's care for his fellow human beings and undergirded by a sturdy faith in the religion of his fathers.

Of all the stretches of the river this site most insistently invites me to move back in time; it is the stepping off place to the legendary past of my boyhood imagination. It is also a convenient departure to the past as presented by the scholars who have told a story truer than my legends, a story that eventually leads up to Raymond Roy, Dr. O'Brien, and the city of mills as I came to know it. The human history of the land now occupied by the cities of Winooski and Burlington stretches back, as I have noted, to the ancestors of the Abenaki at least 10,000 years ago. The story now, however, begins with the first Europeans, briefly surveys the growth of the textile industry, and then settles on the middle years of the twentieth century—Winooski as remembered by two loyal sons. The river is the great natural presence without which the story would have been utterly different.

I grew up with the impression that the eighteenth-century Vermont hero Ethan Allen was little short of Moses and George Washington as a savior of his people. I was led to believe that through noble patriotism, courage, and physical strength, he fought off the greedy Yorkers who would have stolen the Vermonters' land. Here was a man who could bite off nail heads and toss hundred-pound bags of salt over his shoulder with his teeth. A more shadowy figure was his brother Ira, the businessman, political wise man, and founder of the University of Vermont. Of the other brothers and sisters—Heman, Lydia, Heber, Levi, Lucy, and Zimri—I heard little or nothing. The child's picture was of course exaggerated and one-sided but not altogether false. As one historian has written of Ethan Allen,

> Controversial then and now, a demigod to some and to others an unconscionable thug, he is generally conceded to have played, for fair or foul, a dominant role in the affairs of early Vermont and the American Republic. It is, in fact, very

probable that without him Vermont would never have come into being, or, once under way, could not have survived its early years of crisis. It may also be true that by capturing Fort Ticonderoga and securing control over the Champlain waterways during the early years of the Revolution, he saved the new nation from being stillborn.[1]

And he was uncommonly strong and capable of great endurance to which his survival of nearly three years of British captivity, some of it wretchedly harsh, stands as testimony. Strength mattered in frontier conditions, but more important, he had the virtues and vices of a natural leader: persuasive and fluent with tongue and pen, confident to the point of arrogance, impetuous to the point of recklessness, ardent in the pursuit of distinction and glory. It was his zeal for great deeds that led both to the successful attack on Ticonderoga and to his unauthorized attempt to take Montreal, a humiliating failure that made him a prisoner. As a young man and occasionally later, he would resort to his fists to settle a point of rights and honor, and he was never reluctant to use force or, more often, the threat of it, to rally the timid settlers, inspire the bolder ones, and intimidate the enemy. Of one New York official he said, "we shall make a hell of his house and burn him in it, and every son of a bitch that will take his part."[2] But to force he added considerable political sagacity and a surprising restraint. It is a striking fact that in his years of struggle with New York, including several as a formal or informal leader of the Green Mountain Boys militia, not a man on either side was killed. Bare backs were whipped, New York surveyors and settlers sometimes were driven out, their buildings burned, and Ethan's cousin Remember Baker lost his thumb to a New York sword, but no man died. Ethan's independence of mind also led him to question the Calvinist beliefs of many of his time, and he was later to write a book of natural philosophy called *Reason the Only Oracle of Man,* attacking the Bible as revelation, the notoriety of which led a visiting minister who happened by Ethan's grave to write, "An awful Infidel, one of ye wickedest men that ever walked this guilty globe. I stopped & looked at his grave with a pious horror."[3]

As with any protracted political conflict, the details fill books, but briefly summarized, most of the early Vermont settlers had bought their land from speculators who had received grants from Benning Wentworth, the Provincial Governor of New Hampshire.[4] Wentworth made his first grant in the territory that was to become Vermont in 1749 despite knowing that under royal charters New York had a better claim to the region than did New Hampshire. New York's early passivity led him to make more grants until by the mid-1760s he had conveyed some three million acres of which 65,000 went to himself. Most of the early grantees were speculators, but with the defeat of the French at Quebec in 1759 and the end of the French and Indian War in 1763, the New Hampshire Grants, as the territory was called, became relatively safe and many settlers began to buy smaller plots from the big landholders. New York authorities recognized that their hold on the region was in danger and began to demand fees on pain of eviction. From one point of view, probably the most strictly legal one, the settlers were in possession of stolen goods. From another, the kind of justice that held that the sweat of one's brow gave title, at least the smallholders could assert their right to remain. No one bothered to inquire into the dispossession of the Indians. Among the speculators and, eventually, settlers were the Allen brothers, first in southwestern Vermont and later in the northern Champlain Valley.

Zeal for the common welfare blended with financial self-interest in proportions not easy to sort out. All of the Allen brothers bought and sold land, but while Ethan became increasingly the loudest voice in opposition to New York claims, Ira, Thomas Chittenden, Seth Warner, and Remember Baker—all heroes of my boyhood reading—also worked diligently to defend the interests of the settlers. For a long time the most successful financially was "Ira, the brilliant, the imperturbable, and, alas, the sometimes devious," who "preferred to move in the shadows, making quiet and, if necessary, questionable deals, and in the end getting what he wanted without much fuss or fanfare. Few men were so influential in the affairs of early Vermont. Without Ira, it is unlikely that Ethan's star would have risen so high."[5]

Although I know better, lingering traces of my early imaginings picture the Allens and their allies as agents of some grand historical force, destined to surge north to prepare the way for Burlington, Winooski, the mills, Raymond Roy, Dr. O'Brien, and even Charles Fish in Essex Junction, for I too grew up on land once owned by these founders of Vermont. In 1772, Ira persuaded Ethan, Heman, Zimri, and Remember Baker to form a land development partnership that came to be known as the Onion River Company. Buying from earlier speculators, it was not long before they owned about 100 square miles (64,000 acres) of the best land in the valley. Stretching from the lake upriver, their holdings then or later included most of what is now Burlington, Winooski, Colchester, Essex, Williston, Richmond, Bolton, and Jericho. The company continued to buy but also sold what they could to settlers. By the time the Revolution put a stop to new settlement in 1775, they had sold some 17,000 acres but still owned more than 60,000, worth $297,000.[6]

This was land the river had been shaping for thousands of years, and the river itself was critically important to the region's appeal. It provided water power, transportation between the lake and the falls, and "a diversity of sorts of excellent fish particularly the salmon," as Ethan said in a Connecticut newspaper advertisement.[7] Among the Connecticut group who made the first purchase of the Onion River land was Thomas Chittenden, soon to be a staunch ally of the Allens and later the President of the independent Republic of Vermont for most of its fourteen years (1777–1791), and eventually the first governor of the state when in 1791 Vermont was admitted to the Union.

The first European building of which we have definite knowledge in what is now Winooski, then Colchester, was Ira's 32-by 20-foot structure made of stout hemlock logs and serving as a combination fort, store, and company office.[8] Remember Baker's cabin soon went up, for he was the first man of this group to move north with his family. (There were a few New York settlers in the region and, much earlier, probably French settlers at various places along the lake, but little evidence remains.[9]) The fifty or so settlers in the Onion River lands withdrew in the face of British domination during the Revolutionary War, but with the peace in 1783 nearly all of Vermont saw a rapid influx as settlers, mainly from southern New England, came in search of inexpensive land. It is estimated that Vermont's population grew from 17,384 in 1775 to 86,507 in 1791, an increase of over 397 percent.[10]

Ira Allen did his part, buying more land—he owned some 200,000 acres in northern Vermont at one time—and building grist and sawmills on the river in Winooski, Burlington, and Essex. By the end of the century, however, the Allens' days of direct influence were largely over. Surprisingly, it was the hot-headed Ethan who left politics for a comfortable, quiet life as a writer and farmer on the Intervale (quiet except for a shortage of cash and recurring problems with debt), leaving at his death in 1789 enough land to take care of his family. Ira, the shrewd and crafty businessman, fell heavily in debt (he gave the land for the University of Vermont but never paid all of his pledge). He neglected his lands to pursue a bizarre and unsuccessful scheme to import French muskets to arm an invasion of Canada (some say to create an independent Vermont), and fled Vermont in 1803 to escape imprisonment for debt.[11] Poverty-stricken, he lived in obscurity in Philadelphia until his death in 1814 when he was buried in a pauper's grave. His passing rated only this notice in a Burlington newspaper: "In Philadelphia, on the evening of the 15th ult. General Ira Allen, late of Colchester, Vt. aged 64 years."[12]

Grist mills and sawmills came early to nearly every settlement in Vermont because people needed grain and lumber; Ira's first grist mill on the river was probably built in 1783. But Winooski early on heard the faint creakings of the woolen mills whose noisy successors would dominate its history down to my time and whose echoes still resound. The story is told from a variety of perspectives in the essays and reminiscences of *The Mills at Winooski Falls: Winooski and Burlington, Vermont.* In David J. Blow's essay, "Industrial Growth in Winooski," we are told that in 1802, while Ira Allen was still in town, one Amos Weeks drew on waterpower to run a cloth fulling and finishing operation on the Burlington side of the river.[13] In the next few years, he added carding, napping, shearing, and dyeing. Apparently in those early years Weeks, having received the wool from farmers, would have scoured (washed) it, picked it to loosen the matted material, added grease to reduce the charge of static electricity, and carded it, that is, combed it out to straighten the fibers. He then would have sent it to various households where it would have been spun into yarn and woven into cloth. (One newspaper ad suggests that the wool would have come to him picked and greased.) The cloth then would come back to him to be fulled, pounded with power-driven hammers in soapy water to remove the grease and to shrink, thicken, and strengthen it. It would then be napped, brushed with teasel heads (the spiny parts of the teasel plant, resembling thistles), which were set into a wooden frame.[14] The raised nap was then sheared to a uniform height. Dyeing is self-explanatory; depending on the time period, it might take place after scouring, so-called "dyed in the wool," or "dyed in the yarn" or "dyed in the cloth."[15]

A few other small textile enterprises appeared during the next twenty to thirty years, and some of them may have introduced weaving, but it wasn't until the late 1830s that something like a modern integrated factory appeared. It was modern both financially, in that it was built by a group of investors, and industrially, in that all the operations were combined in one system in one building, making it "the first complete woolen factory in the State of Vermont."[16] Vincent Feeney compares its economic and social impact to that of IBM in Essex Junction in the 1950s, another economic stimulus that transformed the area.[17] In its daily work schedule and fluctuating fortunes over the decades, the Burlington Mill Company in Winooski was typical of the many mills that sprang up around New England in the nineteenth century. Among the many difficulties to be faced were lack of capital, the inexperience of workers and managers, the uneven quality of goods produced, the challenges of prudent buying, equipment that quickly became outmoded in a time of growing industrial sophistication, and foreign competition resulting from a superior product and varying tariff regulations. In time through various financial reorganizations and changes of ownership, some of these difficulties were overcome, high-quality cloth was produced, and the operation enjoyed periods of success. War has always been good for the textile business, so the Mexican War and, especially, the Civil War were boom times. We are told that the owner in the latter conflict "turned the entire production of the factory to army blue and in the first ten months of the war showed a profit of $375,000. Even by national standards the woolen mill was becoming a big business."[18] Feeney describes the fifteen years following the start of the Civil War in 1861 as Winooski's "Golden Age."[19]

The experience of this mill occurred, of course, in the context of the growing industrialization of New England and other parts of the nation.[20] In Jennie G. Versteeg's "Counting Sheep and Other Worldly Goods," for example, I learned that Winooski's integrated mill was anticipated by the country's first, Francis Cabot Lowell's in Waltham, Massachusetts, in 1815. Cotton and woolen manufacturing technology was imported to the United States by industrial spies despite British laws to prevent it. From about 1800 to mid-century, the country's major growth industries were textiles (cotton and woolen) and iron. It is hard for us to imagine the profound social and economic consequences of the shift from small-scale, home-centered enterprises such as farming or the crafts and trades to the factory system and an ever-widening market economy. But before I come to the radical transformation of Winooski by these forces, I should mention Vermont's sheep boom, which touched my own family at the time of my great-great-grandfather, a farmer who built the doors in his house extra wide so that bundles of wool could be carried to the attic to be stored for eventual sale.

Historians tell of the great changes in Vermont's agricultural economy during the course of the nineteenth century from wheat to sheep to dairy cattle but with much overlapping of these enterprises and much diversification, for changes of this sort do not start and stop with textbook neatness. At its peak, investment in sheep reminds me of the Dutch tulip mania of the seventeenth century or the Silicon Valley boom of the day before yesterday. Recently imported Merino sheep were so prized for their wool that by 1809, we are told, a ram might bring as much as $1,500. Sometimes markets work fast. By 1815, imports by a number of people, including Colonel William Jarvis of Weathersfield, Vermont, had lowered prices until a ram would sell for just $1. Vermont's sheep population rose over the next few decades to peak at over six million in 1845, these Merinos often crossed with other breeds of sheep to increase the yield of wool. Five years later, the number had plummeted to a little over one million and by the end of the century it had fallen to just under 300,000. The drop in the Vermont flocks did not result from less demand for wool but from cheaper fleece imported from the Midwest and abroad.

The Burlington Mill Company rode the economic roller coaster until 1898 when it went off the track for good. Bankrupt and in receivership, it was purchased in 1901 by the American Woolen Company, which modernized and expanded the operations and ran the business until 1954, the year I graduated from high school, when it closed for good, victim of antiquated equipment, the growing use of synthetics, foreign competition, and domestic competition from the South with its lower costs for labor (non-union), electricity, heating, and construction.[21]

It was the last twenty years or so of the mill's productive life that are so vividly remembered by surviving mill workers or their families. I cannot begin to do justice to the richness of their memories. It is the stock in trade of the progressive view of industrial history to see the mills as dens of oppression, forcing workers, many of them children with no decent alternatives, to work long hours at low pay in miserable conditions. Depending on the time and the mill, there is much truth in this view; some of Lewis Hines' photographs of child laborers were taken in Winooski in the early 1900s. And lifelong Winooski resident Rita Martel told me that her mother, a weaver, said, "I'll do everything in my power so you won't work in the mill." This mother, when she was a girl of twelve (in 1912), was sent by *her* mother to work there, her dress stuffed to make her appear old enough to avoid the child labor laws.[22] Her age discovered, she would be sent home whereupon her mother would redesign her dress and hair and send her back.

But if you listen with an open mind to the reminiscences of the Winooski workers or of the Amoskeag mill hands in Manchester, New Hampshire, a nuanced account emerges.[23] For all of the hardships, mill life often seemed preferable to that on the struggling farms or in the European upheavals from which the workers or their parents had come. As Rita Martel said, "Nine people were interviewed for a video about the mill and there was only one negative comment." The mill could create a fraternal feeling, a community of fellow workers that offered rewards beyond the pay, although the pay itself was often better than could be found elsewhere. And the mill was the economic lifeblood of the community, not just a job but the vital source that sustained everything from barber shops and grocery stores to local government and churches.

It was the mills that brought the ancestors of Raymond Roy and Dr. O'Brien to Winooski along with many other immigrant groups who transformed this once Yankee town into what Rita Martel called a little United Nations. The French have always been the largest group, followed by the Irish and then others including Italians, Poles, Armenians, Syrians, Lebanese, and Jews.[24] And keeping a low profile, often not revealing their identity, there were the Abenaki, frequently bearing French names from the mixed heritage that obscured their Indian features if not, in many cases, their ancestral memories. Raymond's father was part of the French community (Roy is the anglicized Roi), a Canadian immigrant, but he was not a factory hand. He had a grocery store and retail milk route sustained in part by his customers' wages from the mill.

The summer he was sixteen, Raymond worked at American Woolen, and after a year at Burlington Business College (now Champlain College) he was hired full-time in the fall of 1937 in the Efficiency Department, where his love of technical and industrial processes began to flourish.[25] "We had an awful lot of incentive work [piece work] going on," he said, "and it was the Efficiency Department, that really set the rates. We did time studies and all that. We had a stroboscope to check the speed of the machines; it was called a strobotach. That was something." Later he became Safety Director for a time and eventually Superintendent of Worsteds (fine wool cloth). Over the years there were serious accidents and "Some people lost fingers and arms, but proportionally, for the amount of machinery we had, very few large accidents. The percentage was very, very low. When I was Safety Director, we never had a serious casualty and that was about four or five years."

That "very, very low" was expressive of the intensity and exuberance of his speech. He had great admiration for knowledge and competence. The previous superintendent who taught him the job and got him "very, very interested" was "very knowledgeable. You can appreciate that as a Russian immigrant and a Jewish man, he had to know his stuff to hold his job. And he was very strict but very, very fair, and very knowledgeable." Eighty-one years old when I met him, Raymond guided me in words through the workings of the mills with all the enthusiasm of a newly discovered passion, yet it was a story he had lived all of his working life, one he had told hundreds of times before, and was to tell a number of times again in the few years left to him.

Of the changes in woolens from the early days, one was the reclaiming of wool from used clothing bought from rag dealers in the big cities, wool cloth but sewed with cotton thread. The carbonizing plant soaked the fabrics in oil of vitriol (sulfuric acid) that would "burn up the vegetable matter, the cotton, but not weaken the wool fibers to any great extent." The cotton would be reduced to carbon particles. At first the acid waste would go directly to the river, but later a neutralizing agent was added. The river was not Raymond's chief concern, but neither was he oblivious to the impact of the mills. Visiting the great factories on the Merrimack in New Hampshire and Massachusetts as he frequently did, he found so much scum on the river that "You didn't have to be God to walk across the waters."

When I learned a few years later that he had died, I found it hard to believe that this lively, curious intelligence, the high energy and buoyant temperament had ceased to be. His wife Noelia—called Noella, which is easier for English speakers—worked in the mill until the first of their five children was born. From then on she took care of the house and the family. When I visited her to offer my condolences, she was still the lively, spry, spirited lady I had met some years before. Raymond was very conscious of his health, she said; *Prevention* magazine continued to arrive in the mail. He did push-ups and used his treadmill every day until an inoperable, benign brain tumor began to incapacitate him. He fought the disease until the very end, insisting on going up to their bedroom even when he had to crawl on all fours. The problem was first noticed in February 2000, and he died at home in June, a few months before their sixtieth wedding anniversary. Noelia took me to the living room where she keeps his picture on a small table, a basket of flowers on the floor and a small bouquet, kept fresh, next to the photograph. She said that she missed him very much. I told her that the house looked as neat as ever. One of her sons helped with the finances, for "I was never able to do any of that, but I can keep house." Short, lean, spry like her husband, she bent over to show me that at eighty-one she could still touch her toes. "I've still got some zip left," she said.

I met Dr. O'Brien quite by chance at Raymond and Noelia's house one morning. He dropped in to say hello and Raymond, ever hospitable, brought out the bottle of port, and we were soon reminiscing about the people we both knew in Essex Junction; he had kind words about my father. Then seventy-six and still practicing, he had repaid his community many times over for the opportunities it had given him, and he had enjoyed for fifty years the honor that small towns bestow on the good priest, the good lawyer, the good doctor. When I visited him some five years later at his lakeside home in Colchester, he no longer practiced, having suffered a stroke, but he was as sharp as ever and glad to talk about his life in Winooski over the decades.

People were often surprised to hear an O'Brien speak French, but French was the language he first learned at home. "They say love is a universal language," he said. "When they married, my mother spoke hardly any English and my father spoke not a word of French. It was a great marriage; they were happily married for about sixty years." Love aside, French was a necessity in serving the French people in Winooski because many of them spoke little or no English. At St. Francis Xavier, the so-called French church, sermons were in French (in English at the Irish church, St. Stephen's), and nearly all business dealings were available in French: grocery stores, hardware store, haberdasheries. To play ball as a boy with the French-Canadian kids, "You had to speak French because that was the only language they could understand." Feeney says that in all of New England in the early twentieth century, only Manchester, New Hampshire, or Woonsocket, Rhode Island, was as "completely French in its institutions."[26]

But it wasn't just a matter of convenience. There was a local chapter of the St. John the Baptist Society whose mission was to preserve the language and the faith. "Their motto," said Dr. O'Brien, "was '*Etes vous membre? Si non, pourquoi?*' Are you a member? If not, why not?" Another of their expressions, one I heard from Dr. O'Brien at Raymond Roy's house, was "*Qui perd la langue perd la foi.*" Who loses the language, loses the faith. More broadly speaking, it was a way of life that many wished to preserve, but it was a way of life with the church at the core. For some, especially in Canada, there was not only the anxiety over preserving a French way of life in New England but fears of what flight to the United States was doing to French Canada. Priests in Quebec were said to encourage large families to build up French strength within a country dominated by the English—*la revanche du berceau,* it was called, the revenge of the cradle—but there was the continuing drain of emigration, some 600,000 between 1860 and 1900.[27] Devout but not dour, Dr. O'Brien's ironic wit found room to play even in the precincts of the faith: "When Jack Kennedy ran, one of my doctor friends said, 'I suppose you voted for Jack Kennedy because you're a Catholic.' I said, 'What the hell do you think I am, some kind of bigot? I didn't vote for Jack Kennedy because *I'm* a Catholic; I voted for Jack Kennedy because *he's* a Catholic.'" And of a much-revered priest for whom he served as altar boy ("Every Catholic kid had to be an altar boy"), "He was quite a guy. He was a speedy man. He would say mass in about twenty minutes."

Dr. O'Brien spoke of his father with deep respect and admiration.[28] Born in 1892 of Irish immigrant parents, John O'Brien dropped out of high school in his sophomore year to support the family after the death of his father. Starting as office boy, he eventually became the top manager of American Woolen in Winooski and in retirement served his city as councilman and representative to the legislature. He saw the mill struggle during the Depression, he saw it prosper, especially in World Wars I and II, and he saw it close in 1954. Although he became general manager, the family lived in a modest duplex about two blocks from the mill, and Dr. O'Brien remembered hearing the clickety-clack of the looms day and night during the war when the mill, then the largest employer in Vermont, ran three shifts and had over 3,000 employees: "We were used to it, but if anybody came to visit and stayed overnight, they couldn't sleep. It was a noisy place." The mill is said to have produced enough material during World War II for 4,841,200 blankets, 4,000,000 overcoats, and 2,300,000 uniforms.[29]

When Dr. O'Brien, himself a generous man, reminisced about his father, it was not his business success that most moved him but his generosity and compassion. He mitigated the hardships of mill life for his employees as much he could, taking care during layoffs to keep at least one member of a family working, forbidding Bob to work there summers since needier people could use the job, and responding with an open hand to requests for help: "Someone would come by at supper time and need money for food, for whatever, and he would say, 'You go to Mr. Roy's store and you tell them Johnny O'Brien sent you.' He was a very, very compassionate man."

Dr. O'Brien remembered that as youngsters, "We made our own fun." There was no Little League; they played baseball in a cornfield. In the winter, they rode their sleds and jack jumpers in the big gully where St. Stephen's parking lot is now, and sometimes they would ride his uncle's bobsled right down the five blocks of Weaver Street, a kid at each corner to stop cars. It was gratitude for this childhood, not an unusual one for small-town Vermont but fondly recalled, that led him to say, "That's the reason I came back to Winooski to practice. I felt I owed the place something."

Dr. O'Brien found his calling when he turned to medicine. After his undergraduate major in biology at Saint Michael's College, he earned his medical degree at the University of Vermont, and for fifty years took care of his patients in Winooski and the surrounding towns. He never stopped making house calls. One day he made twenty-nine, perhaps setting a record for Guinness, he said. House calls told you a lot about a patient. Mary would come into the office wearing her best dress, for "It was like going to church; going to the doctor was a big event in those days. But you go to the home, and you see she has five kids and they're all crying and fussing, and the old man is sitting in front of the television set drinking beer, and you wonder why Mary has dizzy spells?" Not all of the afflicted were sympathetic sufferers like Mary. One time a mother called the doctor to look at her son, "a real bad actor, forty years ahead of his time, really advanced. He had an eight-millimeter camera and there was a whorehouse on North Street in Burlington and he'd go down there and photograph couples having sex. The mother is hysterical. I walk into the room and I can see that he's been dead for quite a while; he's purple, you know. So I say, 'Call Father McGarrity and give him the last rites.'" It turned out he had committed suicide.

As city health officer, he had to deal with recalcitrant landlords, grocery store owners, and a family that kept a flock of geese in their front yard. "The people raised hell about the geese, so I said, 'I'm the health officer. Just like the Pope—he's supreme in matters of faith and morals and I'm supreme in matters of health.'" He chuckled, but he didn't chuckle when he complained about the HMOs, "the worst thing that ever happened to medicine." They dictate "cookbook medicine," fix hospital stays and the like regardless of the individual needs of the patient. He would like to see national health insurance: "It's a crime we don't have it." I was not surprised to hear that he was a lifelong Democrat because "The Democratic Party was the party of the people," and one sign of the changing times, he said, was that Winooski, long a Democratic stronghold and even voting for Al Smith, was now probably evenly split between Republicans and Democrats. "The whole town has changed," he said. "I used to joke that I never had to stop for stop signs because I knew all the cops in town, and now I don't know anybody." Even his beloved Democratic Party was not what it used to be because "Today the Democrats have been taken over by the wacko liberals."

Change is the one constant from the towns of Cabot and Marshfield at the headwaters of the river down to Winooski and other towns near the mouth. The loss of the mill left Winooski with an unemployment rate of 36 percent, but new jobs were in the offing at the University of Vermont, the Mary Fletcher Hospital (now the Fletcher Allen), and, in great numbers, at the new IBM plant in Essex Junction. Winooski was lifted by the rising economy of the area. In the 1980s, two of the old mill buildings, the Woolen Mill and the Champlain Mill, were converted to apartments, stores, and offices. The new Winooski Falls Park linked the two buildings and the Salmon Hole, and made part of the waterfront an attractive place to walk and sit. Although considerably spruced up by these and other improvements and encouraged by the success of some local businesses, the city center still lacked the economic vitality of, say, downtown Burlington. The Champlain Mill, a major player, lost most of its stores. But now there is new hope. The city looks with guarded optimism at a new, major development on 15 acres along the river, a $75 million project providing 400 housing units and 250,000 feet of commercial space.

French Vermont has changed statewide, not just in Winooski. According to one report, nearly 25 percent of Vermonters in 2000 had French or French-Canadian roots, but in the 1990s French speakers declined by 15 percent. In Winooski itself 39 percent of the population is still French or French-Canadian by ancestry, but French speakers had declined 39 percent, even more than statewide, and in 1992 the parishioners of St. Francis Xavier heard the city's last French mass.[30] The city continues to be ethnically diverse with the addition of a number of southeast Asians.

Though sharp at times, Dr. O'Brien was never bitter and he was a most genial host. About to say goodbye to him and his wife Shirley, I discovered that he had another great love along with his devotion to his family, church, medicine, and hometown—jazz. His all-time favorite jazzman was Johnny Hodges and his opinions here were no less emphatic than in matters of medicine and politics, for Hodges was "The all-time greatest alto sax man, even better than the Bird." Dr. O'Brien played sax and clarinet in dance bands while in high school and college, and he went to concerts whenever he could, the highlight being Benny Goodman's local appearance when he had "his best band ever." This might have been sixty years ago, but Dr. O'Brien rattled off the names of every member including Harry James, Gene Krupa, and singer Liltin' Martha Tilton.

As I drove away I was struck by how familiar I found Dr. O'Brien's early community life, yet how different from my own. I had been in or through Winooski a thousand times as a youngster, and I had watched Saturday double-features in the old Strand Theatre, with its chewing gum under the seats and popcorn grease on the floor. The village intimacy of Essex Junction was not so different from Winooski's small-city intimacy. But for all that, there were differences. Yankee and Protestant, we felt that we had always been where we were, at the center of things, with no fear of being displaced or of losing our way of life; in fact, so secure were we that we seldom spoke of even having a "way of life." The Protestant churches were an integral part of the lives of many of us, but there was nothing particularly threatening from which they had to protect us, not even the Devil, for the hell fires of early New England had long since been dampened. We didn't think of our town as an isolated pocket or a beachhead, for we were part of a web of towns and family that not only stretched across Vermont but characterized it, or so we thought.

Over time, the immigrants and their descendents in Winooski, especially the French, gained the political ascendancy, and a few in the early years had been prominent enough in business to rank among the community leaders. For many, however, it was a long time before they lost the sense of being from somewhere else. Nearly all had relatives in Quebec who maintained customs that the Winooski French tried to perpetuate even as their children moved into the wider American society with all its charms, advantages, and temptations. There was much to gain but much to lose, hence the St. John the Baptist Society, the emphasis on language and faith, the devotion to family. Family mattered to my clan, but we didn't think of it as a protective circle or as something that needed protection. Language, church, and family came together in Winooski with a self-conscious intensity that was rare in the Yankee families of my childhood. And it was a matter of social density as well. Although I found the whole Winooski Valley to be rich with interwoven networks, human and otherwise, the French population of Winooski was perhaps the most concentrated instance of close relationships.The big Catholic families tightened the web. The O'Briens had nine children in ten and one-half years, and extended families were so far-reaching that Dr. O'Brien could say, with some deliberate exaggeration, that in the old days just three French-Canadian families—the Mongeons, the Desautels, and the Villemaires—made up about 50 percent of the population.

Finally, I detected a spiritedness, an animation, that wasn't entirely lacking among the Yankees but was hardly our distinguishing trait. I never entirely agreed with a friend's judgment that the dwindling farm communities, the church vestry wedding receptions, and the Sunday afternoon family visits of my youth were monochromatic, but no one ever called them colorful. It's not a matter of grand gestures or public demonstrations. You notice it in a raised eyebrow, in the inflection of a voice, in Noelia Roy at eighty-one showing that she could still touch her toes. I found it in the port shared with Raymond Roy and Dr. O'Brien at eleven o'clock in the morning. Even after dinner my family never had port.

23

To the Lake

"Intervale." This word has not yet found a place in our dictionaries, and there has been much carping about it . . . But we use it, notwithstanding, because it will express our meaning more briefly and intelligibly to the greater part of our readers, than any other we could employ. It may be derived from inter—within, and vallis—a vale or valley; and in its specific signification, it denotes those alluvial flats, lying along the margins of streams, which have been, or occasionally are overflowed in consequence of the rising of the water.

—ZADOCK THOMPSON, *Natural History of Vermont,* 1853

The country my soul delighted in. —IRA ALLEN

Through the abutments of the old railroad bridge and into the lake at 12:31 PM, June 27, 1997. Hardly a cloud in the sky, calm water, beautiful. Adirondacks in full view.

—MY TRAVEL LOG

Here, below the Salmon Hole, I enter the Intervale.[1] The Winooski River never runs straight for long, but nowhere is its sinuous character more striking than in the lazy, looping course from the Salmon Hole to the lake—for the crow 5 miles, 8½ for *The Tub*—winding its way through the Intervale's 3,900 acres, about three-fifths in Colchester, two-fifths in Burlington, with a very small piece in Winooski.

It's a bright, breezy June morning when I lower *The Tub* into the water for the final stretch of my voyage. Soon I pass the fields of the community farm that I visited with Bonnie Acker and her daughter Dia Davis, ardent gardeners and my generous hosts. Each summer they contribute many hours to the farm. Dia started working here years ago when she could barely walk and continues today as a teenager. Bonnie is a farm board member and editor of the newsletter *Bottom Land News*. She has seen the farm grow from its founding in 1990 to the largest community supported agriculture operation in the state, ranking high in the estimation of the advocates of sustainable agriculture and in fact being named the Vermont Sustainable Agriculture Farm of the year in 1999. One indication of its strength of spirit is how long its competent farm managers have served, Andy Jones for eleven years, Erin Hanley for nine, as of 2003. Andy is one of the main educators of the farm's apprentices and of the incubator farmers (people new to farming). Erin loves the community dimension of her work, pleased that "a thousand people will benefit. I'm really lucky—the best farming situation I can imagine. Someone was watching over me."

Bonnie is now engaged in an ambitious project to improve school lunches through the use of local products. She also gave me lunch, which I ate with the guilt of a secret sinner in church, for just as surely as I enjoyed her garden salad dressed with balsamic vinegar and a touch of maple syrup, the grape bread from nearby Shelburne Farms, and the hummus, carrot sticks, and green olives, I also enjoyed my dinner at Winooski's Papa Frank's, where people line up on the sidewalk for a seat and where for $10.14 plus tip I had linguini with hot sausage, garlic bread, wine, tiramisu, and coffee. My beloved valley will be saved when hot sausage and hummus learn to work together.

The Intervale is a surprising place. In the heart of Vermont's most heavily developed area, it is home to farms and one of the richest habitats for waterfowl, mammals, and fish of any stretch in the entire valley, almost certainly for the widest range of species. Its woodland, farmland, and extensive marshes provide ideal habitat for many species of ducks, for wading birds and osprey, for deer, mink, raccoon, coyote, fox, otter, beaver, muskrat, and other animals; even moose and bear have been spotted here, although they were probably not permanent residents. Field surveys in the Burlington Intervale have counted 144 species of birds, 30 species of reptiles and amphibians, and 22 species of mammals.

All of which is not to say that the Intervale is pristine. The Abenaki and their ancestors left a light footprint with their cultivation of squash, corn, and beans. In the Intervale and nearby, archaeologists have discovered stone points, burial items such as copper beads and shells, and other evidence of human occupation, some of it going back 10,000 years. No doubt much more lies buried under the nearly 10 feet of sediment deposited by the river. The first Europeans to make a lasting impression on the land were the Allen brothers and the settlers who followed in their wake in the late eighteenth century. This was, said Ira, the country his soul delighted in, and Ethan's last home was his farm in the Burlington Intervale.

The Intervale was formed over thousands of years as the river deposited sediment on meeting the lake, the lake shore once miles to the east, and over the centuries the river bed has shifted many times, as is evident from the air. With the rapid conversion of many acres to agriculture as the forests were logged off, farming practices here and upstream destabilized stream banks, eroded hillsides, and washed enormous quantities of sediment downstream to the Intervale, accelerating the natural processes of change. It is estimated that in just eighty years of the nineteenth century, agricultural disturbance caused swings in the river channel of 600 to 1,000 feet. Conspicuous evidence of the movement of the channel is Half-Moon Cove, a cut-off meander.

Although some Burlington residents early recognized the Intervale as being unusually rich in natural history—Ethan Allen's widow Fanny and daughter collected plant specimens here—industrial and commercial interests blind to habitat preservation dominated, along with farming, until very recently. From 1866 when railroad tracks crossed the southern edge to the late twentieth century, the Intervale has seen a succession of uses including a slaughter house, municipal dumps, a bypass highway (Route 127, the Beltline), and the wood-fired McNeil Generating Plant. Some of these ventures chewed up natural areas; impeded the drainage patterns of the wetlands, killing sections of the marsh; and added pollutants including some of the capped dump's leachate that manages to escape despite a collection system. While the relative flatness of the Intervale led me to assume that surface disturbances were less destructive than in steeper terrain, I was surprised to learn that the difference here between well-drained soil and year-around wetness is only 5 feet of elevation. Drainage patterns are critical.

Less-organized activities included private dumping of automobiles, tires, appliances, and anything else no longer of use, and the occupation of some areas by indigents known successively from the early nineteenth century as the strolling poor, paupers, tramps, hobos, and the homeless. There were also the "gypsies," as the Abenaki (and some French-Canadians) were called, who for many years in the summer camped in the Intervale and sold baskets and other craft items. Established residents often were made uncomfortable by those outsiders they perceived as a threat, and a number of efforts were made to rid society of them, including Vermont's eugenics program.[2] The Intervale acquired the aura of a dangerous and unsavory place.[3]

Historically, the natural Intervale was saved from complete absorption into the urban complex by its extensive wetlands and the extraordinary richness of its agricultural soils, and in the last thirty years or so it has acquired a phalanx of allies. Almost everywhere along the Winooski and its tributaries I found professionals or eager amateurs devoted to the ecological health of the valley, but the Intervale has attracted an exceptional concentration of them. Here are five parcels owned by the Winooski Valley Park District, including one at the site of the Ethan Allen Homestead and museum where the organization's headquarters are found. Formed in 1972, the District is a public, non-profit corporation with seven member towns in the lower valley and the goal of adding more towns as far up-river as Montpelier. It now owns and/or manages over 1,700 acres with

12 miles of river, pond, and lake shoreline maintained for habitat and floodplain protection and light recreational use. There is support now from the City of Burlington for promoting a mix of agriculture, recreation, and habitat preservation, and city agencies cooperated with Keeping Track (a wildlife monitoring effort, of which more later) helping the Park District launch its Large Mammal Habitat and Corridors project with the goal of providing as many links as possible among habitats for species that have to travel.[4] The Vermont Fish and Wildlife Department owns the Half-Moon Cove parcel and maintains it as a wildlife management area.

And then there is the Intervale Foundation, formed in 1988, whose various undertakings include community gardens, farming start-ups and education, and composting, all with the comprehensive goal of encouraging economically viable, organic, urban agriculture. Over 300 acres of land have been improved by crop rotation, cover crops, and the application of compost from the Foundation's commercial composting operation. The various farms in the Small Farm Incubator Program, which benefit from low land and equipment rentals, plus the much larger Intervale Community Farm, in which over four hundred participating families buy shares, produced over 600,000 pounds of vegetables in 2000 and served not only local families but also the hospital. This is the farm that Bonnie Acker and Dia Davis have served so faithfully. The Foundation's most ambitious current plan is to tap the generating plant's waste heat for 50,000 square feet of new greenhouses.

Will Raap, owner of Gardeners Supply Company located in the Intervale, was the original moving force behind most of this environmentally friendly, small-scale agriculture, but a loyal cadre has built up over time. Devoted to healthy, tasty food for local consumption instead of to long shelf life and shipping toughness, the Foundation is part of a nation-wide, even worldwide, movement of sustainable agriculture ranging from a few acres of tomatoes for nearby families to the Vermont Fresh Network, providing local produce to Vermont restaurants, to the now-famous organic farms supplying the even more famous Chez Panisse, Alice Waters' distinguished restaurant in Berkeley, California. Organic food has made headway, so much so that corporate agriculture is peering over the fence, but it is community-based enterprises that the Foundation hopes to encourage. Speaking of the closed loop of production and waste recycling, Raap has said, "It demonstrates on a citywide basis that there is a continuum, just like in your garden, between waste, compost, and food . . . It demonstrates that we don't need to be at the mercy of industrial agriculture. We are creating models here that we think are exportable."[5]

It remains to be seen, however, how deeply the small-scale, sustainable principles of the Intervale Foundation and its fellows across the country will penetrate the food-producing and food-consuming habits of the nation at large. The Intervale Community Farm aims to keep its prices somewhat below the cost of organically grown food found in local stores, but tons of non-organic food pass through the stores each year at still lower prices. About the quality of the Intervale food there is little doubt and having just bravely chewed some California strawberries with the texture of cardboard, I share the enthusiasm for reconnecting consumers with the source of their food, growing lettuce, corn, and beans—and strawberries—right under their noses and even encouraging them to grow their own or to participate in community-supported gardens. As with almost every aspect of life up and down the valley, however, from Hap Hayward's mill to the Moultroups' farm, local interests must struggle against the political and economic winds that blow from afar (and even environmental winds, literally, that bring the acid rain) and sometimes they succumb to them. It must be said, however, that if we can speak from the river's perspective, outside influences sometimes benefit local interests, for federal environmental laws have stiffened the state's, which in turn have called for local changes.

I move on slowly. The current is very gentle now, hardly perceptible, and no match for the wind that blows in from the lake. I am eager to reach the mouth of the river but a little sad, too. I feel as if something precious is coming to an end, and indeed it is.

I pass a long point bar on the left bank and across the river, at the outside of the bend, several hundred feet of severe erosion, a mass failure as the geologists would say. I pull over at a small sandbar where a pasture comes to the river. I smell cow manure and see cow tracks in the sand. In the river bottom beneath the canoe is another track, a small furrow with a clam at the end of it. It's time now for neither hummus nor hot sausage but for dried fruit and sunflower seeds, and I sit in a water-smoothed hollow of a fallen tree and watch a sandpiper bobbing up and down. This is not the stirring landscape of mountains and gorges, but the inquisitive traveler would not be bored here.

Looking at the sand at my feet, I remember one of the elementary facts of sedimentation: in moving water the heavier particles drop out first. At the water's edge, I rub a handful between my thumb and finger and note its coarse texture. I walk up the bar a few feet and try it again—coarse but less so. Farther inland, I find a graininess, but I no longer think of it as coarse. And so it goes until graininess gives way to silkiness and I think, "Very fine, very smooth."

Somewhere I pass the spot where a fracas between federal revenue agents and potash smugglers led to Vermont's first hanging. Caught in the war between France and England and angered by the British impressment of sailors from American ships, the United States government imposed an embargo in 1807 forbidding trade with foreign nations. Hurt by this interference with their vigorous Canadian trade, some Vermonters, including the owners of the boat the *Black Snake,* continued to trade clandestinely. The *Black Snake* smuggled potash, used in making soap and glass, carrying it down the Winooski to the lake and north to Canada. Revenue officers finally caught up with it on the river as it was returning from a successful run, and in the encounter the smugglers killed three men. One of the smugglers, Samuel Mott, was later convicted of manslaughter and was punished with one hour in the pillory, fifty lashes, and twenty years in prison. Cyrus Dean was found guilty of murder and in November 1808 was hanged in Burlington, a grand public event that is said to have attracted 10,000 people.[6] How many came for the affirmation of religious and community values that public executions of earlier times were said to promote and how many for the thrill of the grisly spectacle it is impossible to say.

I paddle on. Ahead of me I see a great commotion in the water and hear a loud splashing. Soon I am surrounded by big-scaled fish about two feet long, a school of spawning carp thrashing about in carpish ecstasy. The *Pequod* came to mind when I christened *The Tub* and now, smiling at a similar disproportion, I am reminded of Ishmael's sense of wonder as he observed the amours of great whales in the deep.

The calm of the river is about to be shattered by more than carp. I make the big sweep around Derway Island, passing the high eroding sand banks on the Colchester side. Now I see a few motor boats and then I hear a high-pitched engine something like a chainsaw at full throttle but shriller. As I approach the Delta Park boat access, I see a miniature boat about 3 feet long whizzing across the water. It is powered, I learn, by a weed-trimmer engine and is controlled remotely by an older man in a wheelchair. If I were river czar, how much latitude would I give a crippled man?

The engine's whine recedes as I paddle on. Eventually I round a bend, and ahead of me lies Lake Champlain. I am within yards of where Ethan Allen suffered his fatal seizure returning from South Hero on the ice with a load of hay. No ice today. There's hardly a cloud in the sky, across the lake the Adirondacks rise high and blue, and at 12:31 P.M. I pass through the abutments of the demolished Rutland Railroad bridge and into open water.

24

The Trapper

I know a trapper, a well-educated man, who prefers trapping to hunting because he says that it requires an even greater knowledge of the animal. He is probably right.

—WARNER SHEDD

For a while twenty years ago, full-time trapping and fishing was my only work. I had 400 traps and I never misplaced one; I knew where every one was. It took me from daylight to dark to check them.

—AZARIUS BEAUDOIN

Wildlife professionals recognize that regulated trapping is not a threat to wildlife. The real threat to wildlife populations in Vermont is the destruction and degradation of wildlife habitat.

—VERMONT DEPARTMENT OF FISH AND WILDLIFE, *The Role of Regulated Trapping and Management of Furbearers in Vermont, Present and Future*

The Mouth of the River is more than a geographical location; it is also a neighborhood. When I paddled into the lake and swung south, my destination was the home of Azarius Beaudoin. Azarius and Sylvia live on the lake just south of the river in the Burlington neighborhood with the most colorful and checkered past. The Mouth of the River saw not only potash smugglers like the men of the *Black Snake* but, as the stories go, gamblers, tramps, still operators, rumrunners, shady ladies catering to railroad workers, even pirates. A successful bootlegger still remembered was Joe Guyette, who smuggled liquor down from Canada during Prohibition. According to Azarius, he not only had a fast boat, he knew the lake so well that "He used to go all the way to Canada and back without lights."

Writers, including myself, cannot ignore the dramatic past of this "disorderly cluster of modest houses, camps, shacks, and mobile homes that evolved before modern zoning codes imposed their sterile uniformity." But what appears to the outsider as "a colorful piece of Burlington's cultural history" takes on a different hue viewed from the inside.[1] When I asked Azarius what people called his neighborhood, he said, "People used to say it's the Mouth of the River. I say it's North Avenue Extension. They used to call us the River Rats. It got so we didn't like it any more."

Azarius' father was three when his family came down from Quebec. Azarius' maternal grandmother was Indian, his mother half Indian. They were apparently among the many families who felt the opprobrium of being Indian, for Azarius was never told the tribe and has not been able to find out. When he was four, his family moved to the Mouth of the River, where he lived for much of his boyhood. He dropped out of high school at fourteen when his mother died and the family split up. He lived for a while with an aunt and uncle who never treated him like one of the family: "They acted toward me like I was a stranger to them. They used to send me to bed early and they used to have cake and ice cream downstairs and I could hear." At sixteen in the mid-1940s, he went into the Winooski mills and worked there for about two years, eventually running four looms and making over $200 per week, not so much because he was fast but because he was accurate and took good care of his machines. He liked the job not only because of the money—"I was making more money than people who had worked with those machines for fifteen or twenty years"—but also because of the call for precision and, although the looms made a racket, because of the absence of human noise, a foretaste of the meticulous attention to detail and the solitude of his years as a trapper.

Over the decades, he was to have a number of different jobs including soldiering, farming, driving lumber trucks, and, between trapping seasons, working for various construction companies. In 1963, he and Sylvia moved the family from their farm in Underhill to their present location to live in what was originally a camp like so many of the dwellings here. When he retired, eight years before I first met him, he tore down the camp and built the snug raised ranch in whose spotless kitchen we had coffee and doughnuts and talked about trapping and his early years. It was a hard childhood—little money, few conveniences, and a family that broke up early.

"Well, you made a life for yourself," I said, "despite all that. You did all right."

"Well, I tried."

"When you look back on it, it must look like a pretty good life after that rough beginning."

"No, it's been rough right along. You take and raise six kids and not make big money. . . . But I held the family together, raised them all up, and put them all through high school except one. I said if I didn't go through high school, at least they're going to go through."

So this for me is the Mouth of the River, an unexpected discovery, a comfortable home on the lakefront, a view of the Adirondacks, a hardworking man and woman in their retirement, quietly proud of having come through. What I had set out to find, however, was a trapper, for trapping is as integral a part of the story of human engagement with the valley, and of the whole human occupation of the continent, as hunting, fishing, or hydropower, even if its present scope is much restricted. When Azarius' name was mentioned to me and I first called him, he immediately wanted to know if I was one of the anti-trapping people, the antis as some trappers call them. I assured him that I was not. Anti-trapping voices are heard loud and clear in public debate, and modifications of equipment and methods have occurred partly in response; in some states trapping has been nearly eliminated. Whether elimination has been good for wildlife populations and for human-animal interaction is a matter of debate, which I will touch on later. As with hydropower and dams, however, also a subject of controversy, my purpose is not to enter the fray but to look at the world through new eyes.

When I was a boy, I wanted not only to trap but to become a taxidermist, so I signed up for a correspondence course from the Northwestern School of Taxidermy in Omaha, Nebraska, and for one dollar each month I received a new set of instructions. Memory dims, but I think I gave up after a botched attempt to stuff a bird and after trapping a skunk and bringing it home to the dismay of the neighborhood. When Azarius decided he wanted to trap, he took his father's practical advice and started with house rats: "There were so damn many rats, I used to trap them in the cellar, and skin and stretch them for practice." From house rats he moved on to muskrats and skunks. Skunks got him thrown out of school more than once, and on one occasion he and his friends had a whole section of the Flynn Theater all to themselves.

By the time I first sat in his kitchen, he had over sixty years of experience on trap lines and there wasn't much I could ask that he couldn't answer. Back when he trapped full time, he would often take eight hundred to a thousand muskrats in a single season—his record was twelve hundred. One year he caught eleven otters, fifty raccoons, five minks, and another year twenty-six foxes. (He said that he's not that good at mink.) Twenty years ago, he set over four hundred traps and never misplaced a one, but he would lose an average of four dozen a year to anti-trapping people and, occasionally, to other trappers. He would walk his trap line from six in the morning to five at night and then do all the skinning and fleshing; sometimes Sylvia would walk the line with him. He had little sleep.

Two friends dropped in for coffee, and the talk flowed about trapping, hunting, and eating wild game. Raccoon and bear are good, but avoid the fat, for "It will make you run." Myron Farnsworth, son of the racecar driver Rollover Farnsworth, knew a family that had skunk for Sunday dinner and "It was more tender than chicken." The talk shifted to duck hunting, and Azarius explained why he gave it up. He used to put out decoys in the lake right in front of his house, but one day while waiting in the weeds for ducks to come by, "I heard bang, bang, bang, and my decoys went tumbling end over end. 'What the hell are you doing?' I said, and these little voices said, 'We're shooting ducks.' It was kids eight or ten years old. They had shotguns taller than they were." Confronted with hunters this young, who couldn't distinguish decoys from ducks, who knows what would happen next? He decided it was time to quit. I asked more about trapping.

First you must have a place, and "There are lots of places I used to trap where I can't any more." Over the years new houses, dogs and cats on the loose, and the preserves of the Winooski Valley Park District have closed out a lot of territory. The Park District forbids trapping, but by law you can trap along the river if you stay within 5 feet of the edge of the water. Myron said that he was afraid the District's walking trails would drive away wildlife. Azarius has longstanding arrangements with various property owners (by law he must have permission), some for almost forty years, with written agreements giving him the exclusive right to trap and releasing the property owner from liability if he got hurt. "If they don't want me," he said, "they just have to tell me, but then if they have troubles with nuisance beaver, I don't want to hear about it."

Next you need equipment and supplies, and here, as with the other human pursuits in the valley, ingenuity has produced an astonishing variety. In the more than one hundred pages of *Grimshaws Trapper Supply Catalog* for 2001/2002, you will find Mink Lure No. 33 made from "pure fresh mink urine with several powerful additives, "Fox Lure No. 17 "compounded from the glands of the female fox in heat" with other ingredients, and more than fifty other scents designed to attract animals or mask human odors. You will find pack baskets, skinning knives, snowshoes, trap wax (to eliminate human scent and prevent rusting), fur forms for stretching pelts, boots, gloves, hoes and shovels, fleshing tools (for scraping off fat), fur combs, and of course traps. Traps come in various sizes and three basic types with many variations, only a few of which are represented in the catalogue: cage (live capture, Havahart brand), foothold (live capture, padded or not, long spring or coil spring), and Conibear (body-gripping with quick death, named for its inventor, the Canadian Frank Conibear). (There are also box traps and snares, but snares aren't legal in Vermont.) Prices vary: $5.75 for a small foothold trap, $10 for a padded "soft catch" model, $5.24 for a small Conibear, $20.95 for a big one suitable for beaver, all cheaper by the dozen. Paul Grimshaw also offers about three hundred how-to pamphlets, trap collectors' guides including Robert Vance's *Price Guide to Collector Steel Traps* listing two thousand items, and many books with titles such as *The Mink Manual, Trapping the Fisher, On the Prowl for Bobcats,* and *Coon Come Easy.*

Azarius explained that putting out a trap is called making a set, and the set varies according to the animal. For muskrats, he places a Conibear trap under water in a run with a stick across the run so that the muskrat will duck down and into the trap; it's illegal to trap right at a hut or entrance to a burrow. He doesn't use bait and he doesn't have to worry about his own scent since the trap is under water. An otter set is similar, a Conibear in a run with a stick or grass over the top or under water in an old beaver run. Again, no bait is used, for otters eat fresh fish only. Areas they frequent are easy to spot because of fish scales in the scat. Raccoon habitat varies depending on the availability of food, but they frequent streams and wet areas where you will find their runs. Azarius will put a fresh muskrat carcass near the run with a Conibear trap on either side of the run; the traps can be right on the surface. For beaver, he uses a big Conibear, which must be set at least half under water so that a dog won't wander into it. In fact, by law any body-gripping trap with a spread of more than 8 inches must be either in water or in a tree at least 5 feet above the ground (for fisher).

For the fox, it's handy to have a path, but you don't need one. What you do need is great care in making the set, for the fox is the cleverest of the furbearers (perhaps equaled by the coyote). I remembered what I had learned about the canines' sense of smell, and Azarius confirmed that it is indeed the fox's keen nose that puts the trapper to the test. He looks for a hole in a field where a fox has dug for a meadow mouse or a mole. He prefers an open site rather than a run in the brush because then he can check it from 50 feet away with less chance of contaminating it with his own scent. For bait, he uses some muskrat fat or part of a carcass that he has left in the open for a couple of days so that it has started to decay and has a scent but is not too far gone. He handles it with a stick and covers it with grass. Wearing rubber gloves, he puts some small sticks at one side of the hole, which will encourage the animal to approach from the other side where the trap is lightly covered. Azarius walks around the site as little as possible, sprays the area with fox urine from a mist bottle to mask any scent he may have left, and then retraces his steps to get away. Fox and coyotes usually avoid Conibear and box traps. Azarius thinks it is because they smell the metal; Kim Royar suspects that they are simply too wary to walk into any strange contraption, and she points out that a Conibear big enough for coyotes would not be legal on the ground.

A trapper must check his traps and remove animals every seventy-two hours for traps under water or ice, every twenty-four hours for all others. When he gets home after walking the line, his work is not over. Now he must prepare the pelts for sale. To skin a muskrat, Azarius slits the hide along the back legs to the tail, pulls it down to the head, and cuts out the ears and around the eyes. Next he pulls the skin over a wooden form, scrapes off the fat, and stretches the skin over a wire frame to dry. Once dry, hides can be kept for a long time. Skinning a fox is a lot harder. Azarius hangs up the carcass by the back legs, slits the legs to the tail, then splits the tail to the very tip. Very thin, the hide must be peeled away carefully, using the knife here and there with a delicate touch; the smallest tear will stretch out to a big hole. Since there is little fat, not much scraping is required, and after ten hours on the wire stretcher with the fur inside, Azarius turns it to expose the fur so that a dealer can check its quality. If you don't give it ten hours to dry, the skin may stick to the wire and tear. Since muskrat pelts tend to be all the same, they don't have to be turned. Otters, fishers, and raccoons are skinned pretty much the same way, but their hide is a lot tougher, which is fortunate because it comes off hard. To scrape the fat off raccoon hides, Azarius uses a fleshing bench and a fleshing tool, a scraper that looks much like a carpenter's draw shave.

Beavers are handled differently. Azarius cuts the back legs to the tail, then slits the hide from tail to chin. He removes the paws and pulls the skin over the legs. It is now opened up, flat and round. The hide is tough, there is a lot of fat, and it comes off hard; Azarius uses the fleshing bench. To dry the skin, he tacks it on a piece of plywood. The hides can be big too. The biggest beaver he ever caught probably weighed 75 or 80 pounds, and the hide measured out at 88 inches, the sum of the horizontal and vertical measurements; dealers pay according to size and quality.

Azarius used to sell his pelts to Paul Grimshaw who, in addition to being a trapper and retailer of equipment and supplies, is also a fur dealer. Now he sells to North American Fur Auctions in Wisconsin through an agent here in Vermont. The market is set by forces beyond Azarius' control and the dealer's as well, so there is little room to dicker.

Curious about another avenue of sale, the fur auction, I showed up at the Whitcomb High School in Bethel one Saturday morning for the December auction of the Vermont Trappers Association. Furs were laid out on the gym floor tagged with the trapper's name and other information including whether the furs came from Vermont or out of state. The late Mike Corbin, president of the association, presided over the event with help from his associates. Four dealers were present including the veteran Bill Spears who, with his wife, measured and inspected the pelts for quality. Also present was Kim Royar, Fish and Wildlife biologist and chairperson of the department's Furbearer Project, well regarded by the trappers because unlike some previous biologists she seeks their opinion as to animal numbers and health: "She recognized that we're the ones out there seeing this stuff and seeing what's going on, and she would ask our opinion."

The four dealers sat at tables at the end of the room facing Ron Wright, the auctioneer, whose strong voice and rattling patter generated the excitement auctions are supposed to have. As the trappers presented their lots, the dealers would consult their notes and bid accordingly. Trappers could refuse to sell. Fifty-one raccoons sold at an average price of $11.03, twenty-four otters at $62.60, forty-two minks at $11.70, forty-three red foxes at $30.44, seventy-seven coyotes at $24.95, four fishers at $23.13, and seventy-seven beavers at $10.18. The lowly opossums, three of them, went for an average of $2.00. Mink is a prized fur, but ranch-raised minks hold down the price. When you consider that it can take an hour to prepare a beaver pelt, to say nothing of the effort in the field, $10 isn't much of a return; one man refused a high bid of $150 for twenty-five hides, just $6 a piece.

The market for furs can shift dramatically. At the spring auction in 2000, for example, a little less than two years earlier, the high bid for red fox was $15, the high for coyote $13. From 1997 to 1998, muskrats dropped from an average of almost $3 to just $1.12. Obviously it's the market, but where and what is this market? Mike Corbin and Bruce Baroffio, a director of the association, said that whereas years ago dealers might have sold to manufacturers in the United States, now most of the manufacturers are in Russia, China, and South Korea where labor costs are lower and where a lot of finished garments are also purchased. Dealers at the Vermont auctions probably will send their furs to the big international auctions in Canada, including that of the Hudson's Bay Company, where buyers from around the world "can see 20,000 raccoons and sort them through for color or whatever they want." As with Hap Hayward whose chair rocker business was closed down by Japanese competition for the wood he needed, events in far-flung places can affect what a trapper in the Winooski Valley gets for his efforts or even if he will find it worthwhile to go to the bother. Regarding the events of 1998, one article explained that "the lack of Chinese hard currency and sky-rocketing Russian inflation translate to lower fur prices for Vermont trappers this winter."[2] Back in the 1970s, Mike said, prices were up and many more people trapped. Raccoons would go for $35 to $40, muskrats $5 to $8, and fisher, just then being trapped for the first time since their reintroduction, $300 to $400 compared with perhaps $30 now.

Mike and Bruce lamented the loss of Park District land to trapping, but Bruce said that in general it's easier now than before to gain access for trapping because when people from out of state buy country property and then lose cats to fishers or have their roads flooded by beaver, "All of a sudden they realize they need you." Four people had called Mike for help in the past week alone.

When I heard the name Hudson's Bay Company, images flashed into my mind of hardy trappers and traders in the wilds of Canada. The images were true, but they were only the romantic surface of a complex, world-changing story of exploration, exploitation, business enterprise, and political struggle contributing to the shaping of the United States and Canada as we know them, a story much too complicated for more than a brushing here. Hudson's Bay Company was just one of a number of fur trading organizations—the Northwest Company and John Jacob Astor's companies were among its competitors at various times—but HBC captures my imagination because it is still in business as the world's oldest chartered trading company, having received a royal charter from King Charles II in 1670. It is now a more diversified enterprise that is in fact the fifth-largest employer in Canada, operating more than 500 department stores and employing 70,000 people. With the exception of two brief periods, records of every year of the company's history have been preserved as one of the world's most remarkable business archives, donated by the company to the Province of Manitoba in 1994 in what the Province justifiably calls "one of the most extraordinary public gifts ever." Appraised at almost $60 million, the 2 kilometers of documents are of inestimable value to scholars.[3]

HBC started with furs, but for thousands of years before European contact the native peoples of North America had used furbearers for a variety of purposes, not only for clothing but also for bedding, shelter, boats, binding, ornamentation, religious ceremonies, and meat. They used various kinds of traps including deadfalls, trapdoors, pits, snares, and nets. They were also traders long before they ever saw a European, and although fur remains are exceedingly rare because fur decomposes so quickly, metal, shell, and rock items have enabled archaeologists to map ancient trade routes, some items traveling 1,000 miles from their source, and it is reasonable to conjecture that furs were among the valuable items traded. It is unlikely, however, that the native peoples saw hides simply as units of commercial value. As noted in the Abenaki chapter, animal remains had to be treated in a certain way to show respect, to ask forgiveness, or to ensure that the spirit would return in the form of another animal.

Early as they were in the trade, the HBC men were not the first to bring furs back to Europe. European cod fishermen in the coastal waters of North America purchased furs from the Indians as a sideline from around 1500 to 1580. But in the mid-sixteenth century, felt hats became all the rage in Europe, and odd as it seems, "from about 1550 to 1850 the felt hat industry became the primary driving force behind the fur trade."[4] Other pelts were also traded, but beaver was the backbone supporting the rest. Like a number of other animals, beavers have an outer layer of coarse guard hairs and an undercoat of denser hair. The beaver's underwool is extraordinarily dense, a coat of interlocking hair that actually traps air next to the body even when the animal is underwater, and it was used to make a very fine felt.

That sartorial fashion should have had such far-reaching consequences for the fate of a continent is an argument for the profound irrationality of human history. Alice Outwater has given us a handy synopsis in her chapter "The Fur Trade." Colonists, settlers, professional traders, and their Indian suppliers (later there were white trappers too) pushed ever more deeply into the wilderness, wiping out the beaver in many areas and radically transforming both the Indians' way of life and the ecological character of the land. The numbers are staggering: from the Connecticut River Valley a quarter of a million hides by the 1670s, from New York almost two million by 1700. The impact on the Indians was momentous and irreversible. Once introduced to European goods—knives, guns, metal pots, cloth—they quickly became dependent on them, and continued to trap even as their traditional way of life eroded. The land suffered too, for the estimated 200 million beavers in the precontact United States had created a diverse environment of meadows, wetlands, and rich soils, which was badly damaged as beavers disappeared under the pressure of relentless unregulated trapping and the loss of habitat through agriculture and logging: "Stretching over three hundred years, the North American fur trade would alter the physical landscape of the New World as no other trade has done before or since."[5]

Outwater also reminds us that the European passion for furs was a long-standing one, the Middle Ages being in fact the period of the most extensive use by peasant and nobleman alike. We are told that to supply the English royal household, "from 1285 to 1288 . . . King Edward I bought a hundred and twenty thousand squirrel pelts a year." By the end of the next century, most of the European furbearers had been exterminated, and furs were being imported from Russia, which remained the main source for a long time. Russia provided 96 percent of the 323,624 furs imported into England from March through November 1390. But even Russia, immense as it is, could not satisfy the demand for beaver forever, and "by the mid-1500s, only the remote reaches of Siberia and Scandinavia had ponds still abundant in beavers."[6] North America was next, but by the end of the nineteenth century, demand at home and abroad had decimated many species and nearly extirpated the beaver.

There is a path from all this that leads back to the Winooski Valley and Azarius Beaudoin, for the beaver and many other furbearers have made a comeback in North America, although the perhaps 10 million beavers in the United States now are but a remnant of their ancient numbers. We learn from Outwater that we will probably never regain the ample water table of great stretches of the West maintained by the activity of the beaver along with the buffalo and prairie dog, but returning beaver have helped to restore a healthier ecosystem in many places. In the 1820s and 1830s, Hudson's Bay Company imposed quotas and closed seasons in some of its territory, efforts "designed to place the trapping of beavers on a sustained-yield basis for the first time."[7] Its efforts were not enough, but in the course of the twentieth century, especially in the second half, provincial and state governments regulated trapping more systematically, helped to reintroduce furbearing species where restored habitat would support them, required trapper education before licensing, and encouraged the use of more humane traps and methods. Advances in ecology and wildlife biology allowed wildlife management to become a scientifically based profession although one that must operate within certain political and social constraints. In Ontario, for example, there is close cooperation between the Ministry of Natural Resources and the Ontario Trappers Association (established in 1947), which includes a "registered trapline system" and makes the individual trapper a "resource manager" who is "responsible for managing all the furbearers within a well-defined zone."[8] So-called killer traps and padded foothold traps are promoted. A collaboration of the two groups produced the great compendium of scholarly information upon which I have heavily drawn, *Wild Furbearer Management and Conservation in North America*. This is a significant change from the days when, in the memory of one Ontario veteran, "you knew someone was a trapper if he was cross-eyed—one eye skinning the beaver, the other looking down the lake for the game warden." The same man said, speaking in 1985 of Ontario, that "we are harvesting more fur than the Hudson's Bay Company did at its peak and there is still plenty left."[9]

Small Vermont was never as productive as big Ontario and was for a long time more heavily farmed and more bereft of furbearers. But it now has a healthy population of animals thanks to habitat recovery, effective regulations, a cooperative and knowledgeable trapping community, and the good work of the Fish and Wildlife Department, in particular the Furbearer Management Team under the leadership of Kim Royar. Although an attempt to reintroduce the marten did not succeed, perhaps because of competition from fisher and an apparent tendency to roam—one animal was found in Hartford, Connecticut, 100 miles away from its point of release, another in Rangeley, Maine, 150 miles away—all fourteen of the legally trapped species are plentiful throughout the state.[10]

In Vermont as in general throughout the United States and Canada, wildlife is considered a public resource. Like the managers of the deer herd, Kim's team must take into account not only the health of the animal populations and the habitats they depend on but the interests of what managers call the harvesters (hunters, trappers) and nonconsumptive users as well (photographers, observers). In some cases, humans find themselves the prey instead of the predator as when fishers kill their cats, raccoons raid their gardens, or beavers flood their meadows or septic fields. And then there are the voices of the animal welfare people and the animal rights people, the former arguing that animals whether raised, hunted, or trapped should be treated as humanely as possible, the latter arguing at its most extreme that animals have rights comparable to humans and shouldn't be exploited in any manner.

If the health of a species were the only consideration, Kim and her team could focus exclusively on the so-called biological carrying capacity, that is, the number of animals supportable in the long run by a given habitat. Since harvest must be considered, however, the principle of sustainable yield comes into play, that is, how many animals can be taken—or should be taken—at what time of year to maintain a more or less stable population within the carrying capacity for future generations. It is with these two objectives in mind that Kim conducts the carcass processing of various species at the Roxbury lab as described in "The Bobcat Man." The irate landowner may not know it, but when he calls to complain about beaver flooding, he is letting the managers know that the *cultural* carrying capacity of an area has been exceeded; there is a greater animal impact than people are willing to tolerate. In 2000 to 2002, thirty-four beaver baffles were installed under Kim's direction around Vermont. These are overflow pipes designed to conserve valuable wetland habitat while maintaining the beaver ponds at acceptable levels. Wildlife managers today agree with the animal welfare people. Kim and others in Vermont including the trapper Bruce Baroffio have participated in an American and Canadian effort known as Best Management Practices to test many kinds of traps to identify effective traps that meet animal welfare standards. Many modern foothold traps have been designed to prevent serious injury and have been used widely by managers in research and reintroduction projects.[11] There is no way managers can satisfy the animal rights movement, but they urge everyone to consider the following aspects of regulated trapping.

Although the efforts of Best Management Practices are directed toward ensuring the welfare of individuals, managers chiefly consider the role of trapping in the light of their primary concern, which is to protect the species. If the species isn't protected, more individuals will suffer through displacement and starvation or will never come into existence at all. In the absence of all human interference, many animals die from causes other than old age, and death in a trap is often kinder than what nature has in store, whether predation or starvation as teeth and vital powers fail. Trapping is a cost-effective way of controlling excess populations that can damage the habitat and may—the evidence isn't conclusive—increase the spread of diseases such as rabies and mange. The nutria, a voracious, non-native, beaver-like creature, is moving north from the southern states, destroying wetlands as it goes. Controlling the numbers of some furbearers such as foxes and raccoons can help protect certain endangered animals such as the spiny soft-shell turtle and the eggs and young of endangered birds. Baffles and other passive ways of limiting beaver damage can't do the job alone. When wildlife damage becomes widespread, people lose their protective concern and are less likely to be responsible conservationists. Massachusetts' anti-trapping law passed in 1996 (but since modified) led to an increase in the number of beavers from 24,000 to 52,000 in just three years and doubled the number of damage complaints. Information from trappers in the field and from the analysis of carcasses helps managers regulate in the interests of all parties including the animals. Managers also point out that it's the hunters' and trappers' license fees and the federal excise tax they pay when they buy guns, ammunition, and equipment that support most of the wildlife management conservation efforts. It isn't clear what would happen if legislatures had to provide the money out of general funds.[12]

I walked the trap line twice with Azarius, watched him reset his traps after removing muskrats, followed his keen eye as he commented on beaver channels, animal tracks, and the condition of the marsh. We scared up a young bull moose who rose out of the tall grass like a god and ran off with surprising grace and speed, a 12-foot grapevine trailing from his rack as if he had been resting after some revel. Later that day, the Beaudoins and I drove to Paul and Bernice Grimshaw's trapping supply store in Coopersville, New York. Paul Grimshaw was walking his trap line, but I examined the equipment and supplies and spoke with Bernice, who grew up on a farm, anxious to leave because she worked morning to night, then married a trapper and worked morning to night scraping pelts and running the business. I learned that Paul had sold 269 red squirrels to a taxidermist near Lake Placid who mounts and sells them. Azarius bought a pair of insulated boots and a gallon jug of fox urine with, I hoped, a tight cap, for it went in my trunk. On the way back, he showed me two small Champlain islands devastated by cormorants, and we stopped at the state fish hatchery in Grand Isle where trout are raised but where we also saw on display other fish including some sturgeon 5 feet long, not yet full grown.

In addition to the various ecological concerns and public issues surrounding trapping, there is a tradition that is hard to paint in convincing colors to the urban dweller with no experience of it but that attracted my imagination as a boy and has stayed vivid for Azarius and his fellow trappers for a lifetime. It was handed down generation to generation among native peoples and whites from the time when trapping and hunting provided a significant part of their livelihood. It is compounded of a love of the woods and waterways, a deep practical knowledge of animals, a pride in one's skill, and a sense of adventure that elevates the pursuit of game, for all its physical demands, above the labor of mill or field. Although it would be pure fancy to suppose that the native people's animism, their belief in the spirits of the animals they pursued, survives in the trappers of my time except perhaps among some of the natives themselves, there may be a faint analogy in the tendency of people who deal with animals in immediate practical ways independent of science to ascribe human characteristics to them. It takes a high degree of scientific objectivity to so distance oneself as not to see the industriousness and skill of the beaver, the wariness of the fox, the ferocity of the fisher, the playfulness of the otter. Perhaps it was only natural for humans to assume that they and their fellow beings occupied the same dimensions of reality although with different powers and degrees of dignity. It took powerful teachings of a radically different kind to make us deny what we could so plainly see.

25

Learning to See

Following an animal's trail may bring you closer to the animal physically, but, more important, it brings you closer to it in perception. The longer you follow the animal, the deeper you enter into a perceptual relationship with its life.

—PAUL REZENDES, *Tracking and the Art of Seeing: How to Read Animal Tracks and Sign*

Put aside all your human baggage, the bus that was late, etc. What you want to do out there is concentrate. I can't walk and talk and see anything. It isn't until I stop the clamor in my own brain that I begin to pick up on all the subtle things that are out there.

—SUSAN MORSE of Keeping Track, Inc.

The most interesting animal in the woods is Susan Morse. —MY TRAVEL LOG

My journeys in *The Tub* ended as I paddled out of the river into the open water of Lake Champlain. The reflections to which these journeys gave rise have no end, or so I hope. But stories must stop somewhere. I close my story with final reflections on learning to see.

In the same years that human travel has become ever more rapid and the traveler ever more oblivious of his surroundings, the art of tracking has become increasingly popular. Tracking is the antithesis of speedy motion. It is the slow, deliberate, imaginative attention not just to animal tracks in mud or snow but to every sign of an animal's presence from the scent of urine to a nipped bud. "What you want to do out there," said Susan Morse, "is concentrate."

But why do it at all? The answers, I discovered, were to be found in a surprising range of engagement not only with the natural world but with the soul of the tracker. I found no serious tracker whose only aim was to find the animal and then photograph or shoot it. Although the deer hunter Warner Shedd claims no great expertise as a tracker, deer tracks and other sign tell him much about deer habits and habitats that is helpful to him as a hunter. Beyond that, however, is his desire as a naturalist and conservationist to know as much as possible about Vermont's premier game animal. For the bobcat man and houndsman Patrick Soneira, the track of the cat is essential to the pursuit, but he moves through the woods as a lover of poetry moves through a poem, with an eye for every sign of life and meaning. Kim Royar, John Buck, and other biologists need to interpret sign as part of their professional responsibilities as wildlife managers, but their interest isn't limited to the demands of their job. The eminent trackers Susan Morse and Paul Rezendes bring to their work—indeed their calling—as naturalists and conservationists a passion that is life-defining. For all of these people and more whom I have met or whose writings I have read, there is something compelling about the lives of other beings. At root it may not be different from the connection people feel with their pets or other domestic animals, but since the animals here are not domestic, special skills are required to get to know them.

Despite many hours in the woods as a boy, it took but a few outings with professional trackers to show me how much I had missed. The Paul Rezendes field trip at Massachusetts' Quabbin Reservoir led by staffer John McCarter was called "Tracking: Relearning an Ancient Art," indicating that we were getting a glimpse of the common and necessary skills of our predecessors on this land hundreds, even thousands, of years before. We hadn't walked a hundred yards before John noticed some scat in the middle of the woods road. Canines favor paths or trails. At an inch in diameter too big for a fox, and an odor too faint for a domestic dog, it was probably coyote. In other coyote scat we found deer hair, deer because it was coarse and gray, and because it creased sharply when bent between two sticks. Tiny cuts on a white ash tree were the tooth marks of a gray squirrel marking his territory, the dark spot below possibly urine, a scent deposit used by many animals for various purposes, for as John said, "There are a lot of messages wrapped up in scent." A nearby hemlock was heavily defoliated by porcupines, the twigs cut at a 45-degree angle.

Along Quabbin's beach we found sign of coyote, turkey, moose, deer, beaver, raccoon, bobcat, otter, skunk—and fish. Most of the sign was in the form of tracks, but on an old weathered stump we found bobcat scratches, a rare find. "I have spent hundreds of hours in the woods tracking bobcats," John said, "and I have found only five scratching places." In scratching, bobcats leave scent markers from paw glands, they stretch their muscles, they get exercise, and they trim the outside layers of their claws. John could smell cat urine, but he couldn't locate the exact spot. We found a disturbed place where a skunk had dug; a collapsed beaver bank lodge (these bank lodges, also found along rivers including the Winooski, often don't show above ground); shallow basins or redds in the reservoir's sandy bottom where fish had laid their eggs; and an otter scenting area with pine needles scraped up, urine stains, and scat nearby full of fish scales and bones. We climbed a steep, cliffy slope dotted with porcupine dens among the boulders with piles of scat at the entrances. John said that bobcats often follow porcupine trails in this kind of habitat, both for convenience and the chance of finding a porcupine. Bobcats, like fishers although more rarely, will kill porcupines, not by flipping them over to get at the unprotected belly as commonly believed but by slashing at the face until the animal is too weakened and disoriented to resist. Because fishers can rotate their hind feet around backwards, they can descend a tree head first and attack a retreating porcupine while avoiding its swishing tail. Not only must porcupines remain upright on the trunk, they are indifferent climbers, sometimes falling out of trees to their death. Unlike most climbing mammals, they anchor themselves to the trunk with the hind feet and tail while pulling up with the front; bears, for example, do just the opposite.[1]

Tracking in this extended sense is an exercise of the imagination, adding observation in the field to what we learn in the laboratory. It is an attempt to enter as far as possible into the life of an animal, trying to imagine its felt and perceived world. It was with this in mind that John McCarter had us crawl on our hands and knees through dense thickets on the porcupine trails, to see as a porcupine does from a foot off the ground.

Consider sight and smell. We have all seen house cats picking their way through the back yard, eyes looking everywhere. Dogs, in contrast, run here and there, nose to the ground, seemingly aimless but in fact either searching for scents or following them. We humans see as well as or better than some animals, although much less well than others, but our sense of smell is so much weaker than that of many creatures that it is hard to imagine their minute-by-minute experience. In *The Wild Within,* Paul Rezendes tells us that "an average person has five million olfactory cells. With 220 million olfactory cells, a sheepdog can smell 44 times better than we can." It's not surprising that I have never seen a wild bear in Vermont, for "a bear can smell you from a mile away, and it can smell a carcass from several miles."[2]

Rezendes speculates that "deer smell like we see and we see like deer smell." Deer are color-blind and they appear not to see distinct forms as clearly as we do. For humans, "the world is divided up with boundaries, edges, perimeters," a characteristic that Rezendes suspects is explained by the major part thought plays in our way of seeing: "Our thoughts participate in the act of looking. Thought works by compartmentalizing, dividing up the world into things." He suggests that a deer's vision is more like our peripheral vision in which we do not clearly perceive distinct objects but do see movement. Stand still in plain view of a deer and it may notice you, perhaps be curious or even apprehensive, but not see you as a threat unless you move. If it smells you, however, you will see a white flag disappearing into the woods.[3] For most humans, odors have a rather obscure existence unless close and very strong; we are dimly aware of them, but they don't define our immediate environment as they do for some animals.

Nothing has made the scent world of animals as vivid to me as Rezendes' Scent Visualization Exercise. Rezendes reminds us that when we smell something, we are taking scent molecules into our nose, actual particles given off by whatever it is we are smelling. Now visualize, he says, the molecules that you are giving off, "particles of you—who you are." Imagine them as having a color, let's say purple, so that you can see them: "Soon the room in which you sit is filled with the color of your scent; a purple glow clings to your chair." Now imagine that you are walking through the woods. You will smell, if faintly, many things—trees, grasses, flowers, damp earth—and if you were a deer or a bear you would smell many others. Think of each of these scents as having its distinct color or hue. If you look back along your path, you will see "purple blotches everywhere you stepped." You will see that "the whole forest is a kaleidoscope of color, filled with different scents and rich with information about all who have passed here." Through this exercise, "we can gain an understanding of what some animals can 'see' with their noses. In the same way that we see, animals smell. Their sense of smell is that clear, that defined."

Their smelling is not passive; they do not simply receive the various scents as indifferent physical sensations. Many of them carry a message. As Rezendes says, the scents announce such information as "I am!" "My place." "Where are you?" "I'm over here." "It's mating time." "Keep out—private." The urine of some animals smells different if it's mating season. Scent glands add messages to defecating, scratching, digging, scraping, rubbing, nuzzling.[4] If you allow the word "purpose"—and in some sense it is almost impossible to avoid—almost all the characteristic actions of animals appear to be purposive; they communicate.

All this is "tracking." Back in Vermont with Keeping Track founder Susan Morse, I was a guest at two Keeping Track training sessions for volunteers who would go back to their communities to gather information about animal use of certain important habitats. The purpose of the Keeping Track Monitoring Program was—and is—to accumulate data extensive and scientifically accurate enough to inform community decisions about land use. The ultimate concern is one that I found in the thinking of every knowledgeable person in my travels from biologist to sewage plant operator—the health of the habitat for wildlife, fish, and humans. Habitat health for wildlife, however, consists not only of the conditions in one locality but in the character of various localities and their connection one with another. Since for certain species connectivity is all important, Keeping Track hopes to identify both vital living areas and corridors.

Take the moose, "almost extirpated over the past two centuries," Susan said, "by over-hunting and virtual denuding of the landscape coupled with total removal of the beaver and the loss of beaver flowages that are important habitat for moose in the summer." Since the moose are not "area-sensitive, wide-ranging carnivores, typically used as indicator species" (indicating the health of the habitat), they are an add-on to Keeping Track's basic focal species: bobcat, bear, river otter, fisher, and mink. Susan includes it because of the peculiarities of its range, the importance of connectivity, the paucity of data, and the astonishing success of its comeback. With the recovery of the forests, moose returned from a small source population in northern Maine to New Hampshire, Vermont, and even Massachusetts, Connecticut, and the Adirondacks.[5] Since the amazing growth rate—biologists estimate 10 to 15 percent a year—will eventually cause unsustainable habitat damage and conflict with humans, we need to learn more about how the moose are using the land. The return of natural predators such as cougars and wolves would help, as would hunting aimed at moose population needs. In parts of the Green Mountains, exclusive of the Northeast Kingdom, moose have to climb high in the winter to reach the spruce and fir habitat they need for cover and browse, coupled with hardwood browse such as mountain ash, mountain maple, striped maple, viburnum, birch, and elderberry. But come late March, the moose will begin a gradual descent, browsing and barking trees (stripping bark from trunks), heading for the wetlands for early green vegetation. There the cows and calves will stay into fall while the bulls roam the wetlands and nearby ridges. After the rut, near wetlands, cows and bulls make their way back up to the highlands. Since moose must travel, Susan added, "We need to protect everything between their winter and summer habitats or we will have many more accidents, casualties on the roads." The moose aren't the only casualties. Deer are often thrown off to the side. The much taller moose, up to 6 feet at the shoulder, often comes right over the hood and into the windshield, killing the driver.

Or take black bears. How far they range depends on the abundance and diversity of food in their habitat (for males, 5 to 50 square miles, perhaps half that for females), but requirements can change from year to year. Susan took us to a stand of beeches where we saw so-called bear nests, which I had also seen with Patrick Soneira, clusters of branches high in a tree that a bear will pull in and break off for easier feeding. This was prime bear habitat because the nutritious beech nuts help bears put away those 24,000 calories a day (on a good day) as they fatten up for hibernation, but Susan also spoke of a stand of about 200 mature cherry trees nearby, which bears visit in unusual numbers every five to seven years either because there is a bumper crop of cherries or other food isn't available or both. And in early spring when bears emerge from hibernation, having lost 20 to 30 percent of their weight, they head to the wetlands for the "green-up" of sedges and early spring forbs. If these different feeding areas are cut off from one another by development, the bears are in trouble. Since bears and bobcats like to move along ridges and streams as they keep an eye on potential enemies and look for food, and they pass from valley to valley through saddles or draws, it's important not only to maintain this sort of habitat diversity but to keep it free from frequent human disturbance.

Susan believes in hands-on learning—literally. I was asked to climb partway up a small tree to show how a bear might climb and bend it. The bear as climber has some distinct advantages over a human being. There are the claws, of course, but bears also have a specialized muscle attachment from the scapula to the humeral head, a shoulder adaptation that provides greater leverage for climbing and braking power for descending. Lacking this feature, human beings who climb a lot will have shoulder problems no matter how much muscle building they do. Another advantage is that unlike the human foot, the bear's hind foot has the big toe on the outside, the better to grip a curving trunk. Coming to larger trees, we were asked to put our hands where the front claw marks are and one foot on the scratches of a rear foot. We noticed then that the bear's front feet were fairly horizontal as she gripped the tree while the hind feet were vertical. If you imagine the practical physics of the climb, the pattern makes sense as does the fact that the bear maintained a three-point contact at all times.

In addition to the climbing marks, we saw markings that were intended to send various messages. We came to a mark tree with sign made in June three summers ago, probably by a boar (male bear). "I will demonstrate what probably happened," said Susan, as she took a bear skull, Yorick, out of her pack. "He stood with his chest here, like this, and reached around with his paws. He hooked the bark with his upper jaw and made a couple of passes with his lower jaw. Does everybody see that? I can't recommend enough how important it is to be the animal." Susan said that although scientists have not discovered bear scent glands, bears probably don't need them, for given their sebaceous oils their whole body exudes scent. (Even humans can smell bear scent; in his back yard, near his devastated bird feeder, my neighbor smelled the bear before he saw it.) Bear tree bites are "scent chambers," which help to hold the scent molecules. Most scent marking, in fact, such as otter scent mounds, are both visual and olfactory. Bears also rub against trees, leaving hair that carries a "bouquet of scent." Hair was easy to see when Susan held a white Keeping Track data sheet behind it—"the official Keeping Track rubbing fur search technique." Black at first, hair eventually bleaches to reddish blond or straw yellow. Twenty percent of Susan's regular columns for *Northern Woodlands* discuss scent. She says she has spent more hours in field and library studying scent markings than any other professional tracker.

What are bears saying when they bite, claw, and rub? The larger purpose appears to be mutual avoidance, as along travel routes or around high-value food sources; each animal has its individual scent. Scientists believe marking may also send signals between male and female. And there may be other communication we don't understand. In the Adirondacks near dumpsters or where resorts were still feeding bears to attract them for the tourists, Susan saw marked trails leading to the food source like the spokes of a wheel approaching the hub. Another time as men were trying to capture a yearling female who was living too close to housing developments, the bear "stood up with great concern in her eyes and whined like a puppy, and right in front of my eyes she marked a tree, bit and clawed it, saying, 'Please leave me alone.' I tracked her out of there and everywhere she walked, she marked trees, every hundred feet or so."

Here is the general argument for the monitoring of target species as presented by Keeping Track's Mission Statement:

> It is important to document the presence of various area sensitive species and/or wide ranging fauna, including bobcat, black bear, fisher, river otter, and mink. These species are particularly susceptible to decline; their relatively large home range requirements, coupled with low population densities and reproductive rates make them vulnerable to the ill effects of habitat loss or degradation. Insular populations and fragmented habitats severely compromise the long term well-being of these species. Indeed, the documented presence of these Vermont carnivores serves as a vital indicator of the wholeness and relative health of their wild habitats.
>
> These species may be appreciated as "umbrella" species. Their niche requirements overlap with numerous other taxonomic and ecological functions. Protecting adequate habitat for umbrella species on a landscape scale ensures that species and genetic diversity will likewise be protected.

The case is also made for including moose as mentioned above.

Susan's observations and comments covered a wide range of habitat features and animal behavior. We looked at tiny scratches on bark that may have been made by a raccoon. We spotted a gray tree frog that makes the sweet trill we hear in the spring (not the peeper's peep). We were walking through a managed forest. Just two years ago there were loggers and skidders here, but the work was done neatly and selectively. A professional forester herself, Susan notes that slash from cutting can benefit small animals, and if piled up on a small, dry mound with the coarser material on the bottom, it can make a good den for a bear. If put in a dip, on the other hand, water could accumulate and either drown the cubs or kill them by exposure. We learned that bears don't hoot ("Repeat after me—bears don't hoot"); that a shrieking sound like a woman being murdered is not a cougar, it's a porcupine; and that when bears break open bee hives, it is the protein-rich larvae and pupae they are after, not the honey: "Honey is a bonus." They even raid the nests of hornets and yellow jackets. Handling animal scat is very dangerous, we were told, because of various organisms that can kill you, including a certain species of roundworm found in raccoons, which can damage the central nervous system, and before they find out what's wrong with you, you are probably seriously ill. Look at it, try to identify it, but leave it alone unless you think it is from a lynx or, extremely unlikely, from a cougar or eastern timber wolf, in which case collect it carefully in a plastic bag and give it to the Fish and Wildlife Department.

This remark inevitably led to the question of whether cougars (mountain lions) really live in Vermont. Susan agrees with the Fish and Wildlife biologists that individual animals occasionally appear, but for the same reasons they give, she doesn't think there is a viable breeding population. Cougar sign is not hard to find but is rarely found here. There is no record of any cougar having been hit by a car in Vermont, and there is no contiguous source population from which cougars could enter the state. Sightings, if indeed of cougar, are almost certainly of transients from distant habitats or, more likely, pets released by disaffected owners. On the other hand, speaking as a nationally recognized cougar expert, she disagrees with people who say we don't have adequate habitat: "I've tracked cougars in southern California where there are a hell of a lot more people and a hell of a lot less good habitat—cover, food, water, security, space—all those good things. I'm convinced that cougars and even timber wolves could make a living here, certainly in our wilder country, in the Green Mountains and the Northeast Kingdom." She has even named a hill top Cougar Knob "in honor of the cougars who certainly lived here at one time and may again if we can do our job as conservation planners, which is what this Keeping Track program is all about." We could have cougars, wolves, and Canada lynxes, but if habitat isn't protected, "It will be too late in a matter of decades."

Susan's affection for all animals, as well as her professional interest, was apparent in all she said and did, but if any animal had the edge it was probably the bobcat. She spoke at length about its hunting practices, the "sneaking and peeking" that Patrick Soneira described, its scenting with urine and the glands of its face, tail, and feet, and its varied diet "from small rodents to squirrels to turkeys, grouse, porcupines, deer—you name it, a bobcat will eat it." Its cosmopolitan palate is perhaps outdone only by the coyote's, for whereas "the bobcat's favorite food is anything it can catch, the coyote's is anything it can eat including an old shoe." Because of their diet, a diverse habitat is extremely important for bobcats: "You want wetlands for sure, and if you can also plunk in the middle of that mosaic some good, solid, cliffy refugia, some really impossible hardscrabble places that other critters don't have access to except an occasional peregrine falcon, then you've got bobcat heaven on earth."

One such piece of paradise is Bobby Knob, another spot named by Susan, an extensive cliff and talus escarpment with light vegetation overlooking a vast basin. We felt that we were in a holy place, for we were told in hushed tones to approach the edge of the cliff just six at a time and to walk softly and carefully, trying not to crush any plants or leave any trace of our presence. In winter, bobcat signs are abundant here, but Susan never pushes too close, for she considers it not only rude and intrusive but potentially life-threatening because of the bobcat's limited energy budget. Susan's thirty years of field data confirm what many houndsmen have long said—that cliffy terrain provides critical habitat for birthing, warmth, escape, and resting. In fact, she often backtracks a cat since one can learn just as much by seeing where it has been as by where it is going.

The basin is a vast bowl of almost roadless forest. It is the middle part of a forested region that descends south to the Winooski Valley floor and ascends north and east across a number of secondary roads to the heights of the Worcester Range and other parts of the Green Mountains. This is an "amazing resource," as important as any in Vermont but perilously close to the expanding development of the Burlington area. Thanks to the cooperative effort of conservation groups organized by Susan, 10,000 acres have been targeted for protection with 2,500 already conserved.

Susan founded Keeping Track in 1994, served as its Executive Director, and at the time of this writing is its Program and Research Director. The Keeping Track Monitoring Program is not just another casual exercise in wildlife appreciation. It has trained over 1,300 volunteers representing nearly 100 communities in eight states and one Canadian province. Volunteers must participate in six all-day training sessions in the field and one day in the classroom. The training may be spread over six months to a year at the end of which volunteers are prepared to establish transects, discover and interpret sign, and record data. A transect is a specific route through habitats that volunteers follow four times a year, year after year. Keeping Track field materials include data forms, habitat definitions, a guide to photographing tracks and sign, a special ruler, and a glossary. I have before me the list of habitat definitions used by the Colchester team when I joined them on two treks, one in summer, one in winter. There are seventeen types, from northern hardwoods to meadow/field/open; nine "Enduring Physiologic Features" including steep cliffs, saddle, and wetland; and three slope descriptions: flat, moderate, and steep. The data sheet has places to record all of this information plus seventeen kinds of tracks or signs of the focal species.

The Colchester transect lies a bit north of the Winooski Valley on the 693-acre parcel that includes Colchester Pond and is owned by the Winooski Valley Park District. My treks were at the invitation of Fran

Pepperman Taylor who, with her husband Bob, are friends of the present and a link to the past, for their house looks down on the pond where I used to fish as a boy. One day Bob and I hiked around the pond. From the cliffy height to the east we looked down on a dozen turkey vultures circling above the water. The Champlain Lowlands stretched before us, then the lake and the Adirondacks beyond. We could see the water tower at the University of Vermont, the old Rutland Railroad causeway from Colchester Point to South Hero, the road causeway at the Sand Bar—all just names to a stranger but rich with long association to me. I could almost make out where the Winooski enters the lake. The outlet of Colchester Pond was dammed some years after my boyhood visits there, and I was surprised to learn that the enlarged pond had become prime water for big northern pike, monsters compared to the perch and smallmouth bass I used to catch.

Fran's Keeping Track team on our summer hike included Ron Baker, master of the Global Positioning System, and Gary Sturgis, a veterinarian, whose good ear and love of birds enabled him to identify every bird song we heard including a great crested flycatcher, winter wren, scarlet tanager, hermit thrush, and veery. Almost at our feet an ovenbird fluttered off awkwardly as if with a broken wing to lure us away from her nest. A little searching on the ground under the leaves and we found the nest, of a shape you might consider oven-like.

Walking along, the group decided that we were in what Keeping Track's information sheet would call mixed hardwoods. We recalled that Susan Morse said you could distinguish red maple from sugar maple by the sharp, V-shaped indentations of the leaves: V is for vampire, vampires suck blood, blood is red; or sharpness is for a knife, a knife cuts. . . . Susan's name was often invoked as we made our way through the woods, her words of caution much on our minds as we tried to decide whether claw marks on a tree were made by a bear. "There's no point in recording a wooly mammoth," Susan said, "unless you're sure." With every questionable sign, the group argued it out, Ron the enthusiast, Gary the skeptic, and Fran the cautious moderate at the center.

Just as on Susan's training sessions, being attentive enabled us to see things we otherwise would have missed even if they were not items to be recorded. We found matted places in a hayfield where deer had bedded down, the neat horizontal lines of small holes in bark made by the yellow-bellied sapsucker, the animal paths in an abandoned beaver dam perhaps made by muskrats, the nipped jewelweed browsed by deer, the squirrel bites on bark made either to mark or for the sap. And there was scat, some hard for us to identify. The location of some, on a log, suggested mink, but it was green and slimy indicating an herbivore, perhaps a muskrat. We found what was almost certainly fox scat, containing some fine hair and half of a mandible with an incisor and a molar, probably the remains of a meadow vole. I have read that the reproductive power of voles is such that if it were not for foxes and other predators the earth would be blanketed with them.

The group's training came into full play when we came upon a recent moose track. Fran served as scribe, taking the field notes. Measuring the tracks, Ron and Gary found them to be 4½ inches wide and 7 inches long. The stride was about 6 feet 5 inches, the distance between the steps of the two feet on each side; the straddle was a bit less than 1 foot, the lateral distance between the right track and the left track. Ron found our GPS coordinates. And then a bit later we found moose barking some 6 to 7 feet high on a small red maple. Moose and deer both scrape bark with an upward motion because they have incisors only on the bottom jaw, but this was too high to be deer. The two grooves made by the teeth were distinct and clearly visible.

The winter excursion with snow on the ground offered tracks of mink and fisher, two of the focal species, as well as deer, fox, squirrel, and grouse, and a perfect grouse wing print in the snow where the bird had landed and righted its balance with one wing. Although human noses are not very keen, ours were good enough to confirm the opinion that fox urine has a strong smell. Gary laid a tape measure next to the mink and fisher tracks, then took photographs as evidence of the size. Some tracks were hard to read, reminding us of the importance of gait and other evidence as described by Susan Morse and Paul Rezendes. The thermometer read zero when we set out, but it didn't take us long to warm up, and we ate our lunch in the sun on the very cliff top from which Bob Taylor and I had admired the summer landscape of the Champlain Valley, now white and glistening. It is difficult in Vermont to be free of all evidence of the most ubiquitous mammal, and Gary said he would like to live where he could not hear human sounds.

Though not physically present on these treks, Paul Rezendes and Susan Morse were very much with us, the one through his classic field guide *Tracking and the Art of Seeing,* the other through her passionate and determined work with Keeping Track whose mission the Colchester team had embraced. Ardent, knowledgeable, and skilled, and equally moved by their desire to understand our fellow creatures, they came to their callings by quite different paths. There was much in Susan's childhood that pointed to her life's work: happy hours outdoors on her grandfather's big Pennsylvania farm, family work in forestry and landscaping, early encouragement of her interest in wildlife, a family friend who introduced her to tracking.[6] At Pennsylvania State University, she began a major in forestry but found herself drawn to literature and, as one account has it, probably became "the only student forester publishing poems in journals." Transferring to the University of Vermont as a junior, she majored in English and eventually wrote a master's thesis on *King Lear,* but the spirit of a naturalist was still very much alive, moving her to start Project Elm, a university effort to support the battle against Dutch Elm disease and to save the remaining campus elms through various experimental methods.

One day out hiking, she came upon some bobcat tracks and suddenly "seemed to actually 'see' the cat, where it crouched behind a spruce, then advanced and crouched again." Pushing on, she found where the cat caught and ate a snowshoe hare: "Deciphering tracks turned out to be like interpreting a literary passage. Following her lifelong love of the woods, Susan decided to become a self-directed biospheric scholar. Tracks would be her text." Along the way she worked with cougar expert Harley Shaw of the Arizona Game and Fish Department, co-founded a mountain lion track count in Arizona, and became a nationally recognized authority on wild cats, working with groups across the country to monitor the habitat of the big cats and other indicator species. Training volunteers is at the heart of Keeping Track because there isn't enough university and government money in Vermont or elsewhere to do the field work that needs to be done and because it raises public awareness of wildlife and habitat issues. Except as it relates to her work, not much about her personal life emerges in conversations and interviews. I sensed that her spiritual odyssey followed the track of her public career, or perhaps the reverse; the two are inseparable.

Paul Rezendes was born under a different star. His Nature Programs is a business enterprise (now under new ownership) offering various outdoor education programs including Wild Foods Foraging, Camouflage for Kids! (and Parents), Forest Ecology, Pathfinding without Compass & Map, and Tracking Intensive (a six-day workshop). As Paul Rezendes Photography, he sells limited-edition, custom-made prints of natural scenes. But the program that appears to draw most deeply from his experience is called The Wild Within: Tracking the Inner Landscape with Paul Rezendes. Here the focus shifts to tracking "the most cunning animal we may ever encounter—self."[7]

The title of the program is from his book, *The Wild Within: Adventures in Nature and Animal Teachings.* You might think at first that "wild" here refers only to the human potential for disorder and crime, for Rezendes tells us that "In the early 1960s, I was the leader of two notorious motorcycle gangs. I led a violent, criminal life. I dealt drugs and I participated in events that now seem to belong to another life, another person." Arrested for drugs and an illegal firearm and sentenced to a prison term of five to ten years (suspended), he turned his life around, tried and failed to find his way in Catholicism, and eventually blended the practice and insights of yoga with his growing knowledge of "the language of the forest." His ambition has been nothing if not bold: "Nature inspired me to spend a lifetime asking the deep questions that are at the heart of this book: What is thought? Where does the self end and the world begin? What is truth? What gives meaning to our lives?"[8] I cannot do more than hint at his answers, I suspect that academic philosophers and psychologists would look for more of what passes for rigorous method, and Rezendes himself says that answers to questions of this sort are not the kind that can be neatly passed from one person to another. I find his book useful because it seeks to bridge the divide between the world of other animals and our human world of intense preoccupation with self, to break out of the "trance" of "self-centeredness" as Bill McKibben says in the Foreword. Something like this desire to close the gap, if not so explicitly articulated, I have found in a number of the naturalists and outdoorsmen in the Winooski Valley.

So what is this wild within? For Rezendes the web of life, a fundamental natural fact acknowledged by everyone I met on my travels, takes on an extra dimension. It enables him to imagine an eternal now, an ever-changing present in which everything from the past has its consequences in the present and everything in the present will have consequences in the future. Rational thought divides time into past, present, and future just as our vision, in league with thought, divides the physical world into sharply differentiated forms. The notion of an ever-changing, eternal present enables Rezendes to experience the present more fully, to attend to what is within and around him, because he is not always escaping to the past or the future.[9]

In this expanded present, we may learn to distinguish between what Rezendes calls body fear, an "intelligent bodily response to danger," and psychological fear, the anxiety we feel because of threatened vanity, status, self-esteem. It was fear of this sort, he says, that made him work so hard to maintain his image as the courageous leader of motorcycle gangs; when other people accepted this image, he enjoyed "a sense of well-being and power." In our exploration of the inner landscape, we should see that "Fear is an opportunity. Approach it like a tracker in the forest—watch where it goes, what it does, what it eats, where it rests, where it turns, where it stops, where it hides. Embrace it. Stay with it."

The most obvious form of psychological fear is our fear of death. And now Rezendes takes a step—a leap—that many will not be willing to follow but that is, I think, one variant of the recurring human dream of escape from the painful smallness of our individual identity, of hope for union with a larger reality—the mystical marriage with God, the merging into the all, the new dimensions of psychotropic adventures. In his most sweeping formulation, describing a new "seeing," Rezendes says,

"In that moment I saw that wildness was the whole universe. All things were manifesting in it." And even that "I was the universe. . . ."[10] The fear of death, he says, is the fear of losing the self that we have fought so hard to maintain, but

> Glimmering on the edge of death, sweeping in to overwhelm us, is the larger sense of who we are. It is there all the time inside us, but we avoid it with our fear, with the image of who we think we are. This larger self is what I call the wild within, the whole of the conscious universe. In that expansive awareness of our larger identity there is no fear, and if we watch fear from that place—not trying to conquer or control it but just understand it—then fear will start to atrophy. It will diminish and shrivel, perhaps not immediately but inevitably.[11]

This is the concept of the wild within at its most far-reaching. In other formulations it is much closer to Susan Morse's admonition to concentrate. To concentrate in the woods—and no doubt elsewhere—is in a sense to forget yourself in order to attend to the here and now. Rezendes, as usual, gives it a sharper twist, saying that people can "drop their everyday personae" and "*become* the forest."[12] In a number of other passages as well, he links the readily accessible idea with ones that give us pause. We must learn to be aware, to be attentive, but to do so we need to understand how our "thought" tends to work against it. Thought is that remarkable and necessary power of the human mind to imagine, anticipate, remember, categorize, calculate—in short to do many things that take us away from attending to the moment. I might argue that attending is itself a thoughtful activity, but I would concede that it is easily overwhelmed by other thinking processes. I have often driven miles with only the dimmest awareness of the passing landscape.

The words of Loren Eiseley and Wendell Berry (in the epigraphs of chapter 1) suggest one major context in which our relations to the natural world are played out. At one extreme, we contemplate the "tenuous thread of living protoplasm" that stretches through a succession of life forms, many of which exist only as fossils if at all. From this perspective the "forms are phantoms. The thread alone is real; the thread is life." At the other extreme, we contemplate the "uniqueness of an individual creature" whose "life is all that happens to it in its place." Its "wholeness" cannot be captured by scientific objectivity alone: "It shows itself to affection and familiarity."

The naturalists I have come to know, managers and others, know better than most the anonymity and precariousness of individual lives both historically and in the present. They must deal with the thread of life as it manifests itself in populations of the plant and animal kingdoms. But these populations are, to them, more than abstractions to be manipulated. They are living beings entrusted to a wise care that is born of familiarity and affection. The dream of a wilderness untouched by humans is just that, a dream. "Management" is a prosaic term, but disciplined by responsibility and inspired by love, it is a noble calling. We are the caretakers of the planet; we must attend. It is our fate.

In its own meandering way, my book has been an attempt to attend. Passive observation brings many beautiful things to the mind's eye, but since attending is enriched by knowledge, I end where I began, with gratitude to the many people who helped me to see the valley through their eyes. Viewed from a certain distance, human beings also melt into the living protoplasm that is the thread of life. It has been my privilege to spend time with some unique individuals in their place, in one small valley in Vermont.

Notes

Full bibliographic information is given in Published Sources, page 243.

1. My River, My Valley (pp. 1–5)

1. For much of this paragraph, see Meeks, *Vermont's Land and Resources,* 14; Klyza and Trombulak, *The Story of Vermont,* 22–24; Winooski Conservation District Local Work Group, *What Is Your Connection?;* Usinger, *The Life of Rivers and Streams,* 163.

2. Meeks, *Vermont's Land and Resources,* 8; Klyza and Trombulak, *The Story of Vermont,* 34–41; Johnson, *The Nature of Vermont,* 22–32.

2. Headwaters (pp. 6–13)

1. Usinger, *The Life of Rivers and Streams,* 70.

2. See Jan Albers' *Hands on the Land* for a richly illustrated account of the history of the Vermont landscape.

3. Outwater, *Water: A Natural History,* 63 and, generally, chapter 4, "The Voyage of Rainfall." For early, dramatic descriptions of the devastation caused by deforestation, see Marsh, *Man and Nature,* chapter VI, "The Woods."

4. Meeks, *Vermont's Land and Resources,* 183–84.

5. Phillip "Bob" Lamberton, quoted in Pitkin, *Cabot, Vermont,* 36.

6. Apparently not all Cabot men were as eager to serve as those on the monument. About 150 did serve, but 11 paid to avoid service under the legal arrangement whereby "A man could either pay commutation to the government for a substitute to go in his place, or could procure his own substitute, making whatever deal was agreeable to both parties" (Pitkin, *Cabot, Vermont,* 168). Jane McKay said that during the Civil War men had her great-grandfather, a doctor, pull their teeth "because apparently if you didn't have teeth you couldn't go to war because you had to chew hard tack" (ibid., 72).

7. Hausman, "Rock Collector Takes His Hobby Seriously," *The Caledonian Record* (1997).

3. *Growing Up at the Headwaters (pp. 14–21)*

1. Pitkin, *Cabot, Vermont,* 77.
2. Ibid., 114.
3. Ibid., 113.
4. Ibid., 46–52; Pitkin, *Marshfield, Vermont,* 34–38, 84–86.
5. Neill, *Fiery Crosses in the Green Mountains.*
6. Pitkin, *Cabot, Vermont,* 57.
7. Ibid., 56–57. *Cabot, Vermont* also has the story of Eva Kurz who said that "people were not nice to me" when she was married to her first husband, an Italian. Then she married Roman Kurz, a German, and had to endure suspicion during World War II that her husband was a Nazi. In 1943, Roman headed up the War Chest Drive in Cabot, raising over $1,000 (56, 174).
8. Ibid., 23.
9. Ibid., 27.
10. Barbara Carpenter in conversation and Wallach, "Barbara Carpenter," 53–56.
11. Hilt Abbot in conversation and Pitkin, *Cabot, Vermont,* various.
12. Pitkin, *Cabot, Vermont,* 76.
13. Wallach, "Barbara Carpenter," 54.

4. *Poking Around, Poking Along (pp. 22–28)*

1. Cabot Creamery for information and quotations.
2. Pitkin, *Cabot, Vermont,* 19–20.
3. Ibid., 23.
4. Ibid., 2.
5. Pitkin, *Marshfield, Vermont,* 54.
6. Somewhere in my travels I was given a copy of extracts from Cole's log. See Pitkin, *Marshfield, Vermont,* 19, 24, 71, 72, 87.
7. Wessels, *Reading the Forested Landscape,* 123–27.

5. *Hap's Mill (pp. 29–34)*

1. Klyza and Trombulak, *The Story of Vermont,* 101–103.
2. Pitkin, *Marshfield, Vermont,* 55. See 54–55 for more information about power generation at Molly's Falls.
3. Pitkin, *Cabot, Vermont,* 34.
4. Ibid., 33.
5. S. Morgan Smith Company, 7.
6. S. Morgan Smith Company. Mathematically inclined readers might like to know that "the revolutions and water consumption of a turbine vary as the square root of the head ($\sqrt{H}$) and the power varies as the square root of the cube of the head ($\sqrt{H^3}$). This means that when the head is multiplied by four, the speed and water used per minute will be multiplied by two and the power multiplied by eight (15).

6. *The Making of a Valley (pp. 35–46)*

1. Apparently Friedrich Schiller's phrase, quoted by Weber in the following passage and elsewhere: "The fate of our times is characterized by rationalization and intellectualization and, above all, by the 'disenchantment of the world.' Precisely the ultimate and most sublime values have retreated from public life either into the transcendental realm of mystic life or into the brotherliness of direct and personal human relations." Weber, "Science as a Vocation," 51.

2. Discussed in many places. I drew on Lutgens and Tarbuck, *Essentials of Geology;* Dathe, *Fundamentals of Historical Geology;* Klyza and Trombulak, *The Story of Vermont;* Johnson, *The Nature of Vermont;* and Meeks, *Vermont's Land and Resources.*

3. Johnson, *The Nature of Vermont,* 5.

4. Gilluly, Waters, and Woodford, *Principles of Geology,* quotations from 218, 473.

5. For general information, I also have drawn heavily on Lutgens and Tarbuck, *Essentials of Geology,* 377–89, 398–414; and Dathe, *Fundamentals of Historical Geology,* 11–17, 170–77.

6. Klyza and Trombulak, 9–15.

7. Ibid., 13.

8. Meeks, *Vermont's Land and Resources,* calls the fault "one of the best examples in the world of this type of feature" (65).

9. Jacobs, *The Physical Features of Vermont,* 51.

10. Meeks, *Vermont's Land and Resources,* 18–19.

11. Lutgens and Tarbuck, *Essentials of Geology,* 183. This information about the age of rivers follows Lutgens and Tarbuck, 179–83.

12. The story of glaciation has often been told. For a general account, I have relied most heavily on Lutgens and Tarbuck, *Essentials of Geology,* chapter 11. For the impact on Vermont, I have used Johnson, *The Nature of Vermont,* 16–24; Meeks, *Vermont's Land and Resources,* chapter 3; Klyza and Trombulak, *The Story of Vermont,* 15–18; Stewart and MacClintock, *The Surficial Geology and Pleistocene History of Vermont,* 139–84, 224–35; and Larsen, "History of Glacial Lakes," and "Glacial History," 286–300.

13. Imbrie and Imbrie, *Ice Ages,* 87.

14. Recer, "Deepest Ice Age Pushed Earth along Evolution."

15. Lutgens and Tarbuck, *Essentials of Geology,* 386, citing Don L. Eicher.

16. Meeks, *Vermont's Land and Resources,* 56.

17. Howe, "The Charlotte Whale."

18. Thompson, *Natural History of Vermont,* 38, 233.

19. Larsen, "Glacial History."

20. Klyza and Trombulak, *The Story of Vermont,* 23.

21. Lutgens and Tarbuck, *Essentials of Geology,* 209.

7. *I Take to the Water (pp. 47–51)*

1. Roth, *The Complete Beginner's Guide to Canoeing,* 13.

2. Ibid., 11.

8. The Deer Hunter (pp. 52–63)

1. The information in this paragraph is drawn from Flannery, *The Eternal Frontier,* 155–69 and Klyza and Trombulak, *The Story of Vermont,* 23–27.

2. Flannery, *The Eternal Frontier,* 186.

3. Shedd, "Think Deer," *New Hampshire Wildlife* (July–August 1985).

4. This remark and other quotations and information not otherwise attributed are from a conversation with Warner Shedd.

5. Shedd, "Think Deer," *New Hampshire Wildlife* (July–August 1985).

6. Shedd, "Think Deer," *New Hampshire Wildlife* (May–June 1996); (July–August 1996).

7. Flannery, *The Eternal Frontier,* 220.

8. Scanlon, "The Biggest and Best," 20–21.

9. Shedd, "Wildlife Matters," *Vermont Outdoors Magazine* (August 1999): 7.

10. The careful reader may wonder about Warner's reaction when Mark killed his first deer: "I grabbed him in a bear hug and swung him around." The difference, I suspect, is that Warner was happy for his son whereas the offending hunter was exulting over the death of the deer.

11. Shedd, "Wildlife Matters," *Vermont Outdoors Magazine* (November 2000): 14.

12. Nelson, *Heart and Blood,* 22. Deer literature is vast. John Buck, head of Vermont's Deer Management Team, recommended Nelson and Lowell Halls, *White-Tailed Deer.*

13. The information in this and the following paragraph is widely available. I drew on conversation with Warner Shedd; Buck, "The State of Fish and Wildlife" (two articles); Deer Management Team, "Deer Wintering Areas and Development in Vermont"; Shedd, "Wildlife Matters," *Vermont Outdoors Magazine* (March 1998): 6; Smith, "Test of Survival," 30; Hentcy, "The Sporting Naturalist," 28; Bernier, "Maine Whitetail Connection," 19.

14. Nelson, *Heart and Blood,* 45.

15. Ibid., 99.

16. Halls, *White-Tailed Deer,* 60.

17. Nelson, *Heart and Blood,* 104, citing the work of John F. Reiger, *American Sportsmen and the Origins of Conservation* (Norman: University of Oklahoma Press, 1985).

18. Halls, *White-Tailed Deer,* 72.

19. Deer Management Team, Vermont Department of Fish and Wildlife, *White-Tailed Deer Management Plan for the State of Vermont (1997–2006).*

20. Thompson, *Natural History of Vermont,* 51.

21. Halls, *White-Tailed Deer,* 212.

22. Blais, "The Buck Stops Here," 28.

23. Vermont Department of Fish and Wildlife, *A Report of Vermont's 2000 Deer Hunting Seasons; A Report of Vermont's 2001 Deer Hunting Seasons.*

24. Nelson, *Heart and Blood,* 15. See also McCabe and McCabe, "Of Slings and Arrows."

9. *In Search of the Wild Onion (pp. 64–67)*

1. Barg, *The Streamside Sentinel,* 3.

10. *From Influent to Effluent: A Clean Story and Some Dirty Ones (pp. 68–78)*

1. Vermont Water Quality Division, *Vermont Water Quality Standards;* Shedd, "Wildlife Matters," *Outdoors Magazine* (April 2001): 11.

2. The Federation focused on "impaired water," some of which was in fact suitable for fishing and swimming but had pollution that would be harmful in the long run if not corrected. Shedd, "Wildlife Matters," *Vermont Outdoors Magazine* (July 2000): 12. Also see Shedd, "Wildlife Matters," *Outdoors Magazine* (March 2001): 17; (April 2001): 11; (February 2002): 9.

3. Outwater, *Water: A Natural History,* 137.

4. This paragraph and the following are mainly from Usinger, *The Life of Rivers and Streams,* 109ff.

5. "Chlorophyta," *The Columbia Encyclopedia,* 545.

6. Usinger, *The Life of Rivers and Streams,* 115–16.

7. Paul Olander, Director of Vermont's Wastewater Division, provided information about the impact of excessive algae and about the classification of the treatment plant.

8. "Bacteria," *The Columbia Encyclopedia,* 203.

9. Estrin, *Montpelier Rivers Report,* 36–37, gives the relative pollution tolerance of eight fish commonly found in the Winooski and other streams in Montpelier. Most tolerant is the white sucker, intermediate are the longnose dace and the common shiner, and intermediate to intolerant is the longnose sucker, none of them game fish. Joining the slimy sculpin as intolerant are the species most prized by fishermen, the brook, brown, and rainbow trout.

10. Montpelier (VT), City of, "Wastewater Treatment Facilities."

11. Some of the information in this paragraph is from Paul Olander, Director of Vermont's Wastewater Division, and from Stephen Gray, Montpelier's Director of Public Works.

12. Outwater, *Water: A Natural History,* 145.

13. Winooski Conservation District Local Work Group, *What Is Your Connection?,* 3.

14. National Environmental Education and Training Foundation and Roper Starch Worldwide, "Myths and Facts about Water."

15. Vermont Water Quality Division, *Vermont Water Quality Standards.*

16. The following information and quotations are from Outwater, *Water: A Natural History,* 149–63.

17. U.S. PIRG, "Permit to Pollute."

18. Estrin, *Montpelier Rivers Report,* 32.

19. Basilchuk, "Most Lake Pollution Comes from Yards," 1A.

20. Pease, "Setting the Standards," 5.

21. Vermont Department of Environmental Conservation, et al., "Vermont Unified Watershed Assessment."

22. The sediment information in this paragraph is from Estrin, *Montpelier Rivers Report,* 24.

12. Stream Dynamics: The Water Woman and the Surficial Geologist (pp. 82–93)

1. The other tributaries of note are Alder Brook (10 miles, 6,900 acres), Mill Brook (9.5 miles, 10,350 acres), Little River (Waterbury River) (15 miles, 71,070 acres), North Branch (Worcester Branch) (18 miles, 48,300 acres), Kingsbury Branch (12 miles, 34,500 acres), Jail Branch (16 miles, 31,050 acres), Stevens Branch (13 miles, 82,800 acres), Dog River (18 miles, 60,030 acres), Huntington River (20 miles, 43,470 acres), Muddy Brook (5.5 miles, 16,560 acres). Winooski Conservation District Local Work Group, *What Is Your Connection?,* 2.

2. Friends of the Mad River, "Friends of the Mad River."

3. As quoted by Faulkner, *Francis Bacon,* 252.

4. Looking at the long run, some writers question the idea of balance and harmony, arguing that even independently of human influence we find "chaotic dynamics in systems, and long-term disequilibrium and flux" (Krech, *The Ecological Indian,* 22–23).

5. Gow, "Stabilizing the Streambank," 1.

6. Vermont Agency of Natural Resources, *Stream Geomorphic Assessment.*

7. Barg, *The Streamside Sentinel,* 4–5. This chapter relies heavily on this publication.

8. Barg, *The Streamside Sentinel,* 7.

9. Vermont Agency of Natural Resources, "Streambank and Lakeshore Vegetation Management Procedure," 5.

10. Quinlan, Ryan, and Shepherd, *Vermont Stream Bank Restoration Guide.*

11. Barg and Springston, "Floods and the Great Brook," 16.

12. Ibid.

13. Ibid, 19.

14. Ibid., 16.

15. Ibid., 19.

16. Vermont Department of Environmental Conservation, *Fluvial Geomorphology,* 1.

17. Knighton, *Fluvial Forms and Processes,* 28.

18. Ibid., 113.

13. A Perilous Portage in a City of Rivers (pp. 94–98)

1. Dwight, *Travels in New England and New York,* vol. 2, 304. Dwight, who was president of Yale from 1795 to 1817, said that in such a small town, legislators would be restricted to "confined and degrading conversation" and "trifling and contemptible" amusements.

2. Estrin, *Montpelier Rivers Report,* 12.
3. Ibid., 27.
4. Ibid., 1.
5. Ibid., 37.
6. Ibid., 3.

14. The Trout Fisherman (pp. 99–112)

1. J. M. Martin, *Vermont Stream Survey 1952–1960,* 81.
2. Cousins, "Opportunity for Change," 17.
3. Lange, *Wildlife Journal.*
4. Holland, *The Trout Fisherman's Bible,* 71.
5. Quick, *Trout Fishing and Trout Flies,* 25–35.
6. Cairns, "Flies on the Water," *Outdoors Magazine* (July 2002): 15.
7. Schwiebert, "The Evolution of Fly Fishing," 1–7; Keating, *Trout Fishing,* 54.
8. Schwiebert, "The Evolution of Fly Fishing," 6–7.
9. Keating, *Trout Fishing,* 56.
10. Cairns, "Flies on the Water," *Vermont Outdoors Magazine* (August 1999): 23.
11. Keating, *Trout Fishing,* 56–58.
12. Guertin, web site, "Flies for Vermont Trout Streams."
13. Pyne, *Outdoor Journal,* November 3, 2002.
14. Dick, *The Art and Science of Fly Fishing,* 46.
15. Blaisdell, *The Philosophical Fisherman,* 88, 70. Lawrence Pyne calls Blaisdell "the best outdoor writer to have ever called Vermont home" and this book "a must-read for every Vermont angler who loves his sport and his state." (Pyne, "Lake Champlain Valley Report," 28.)
16. Cairns, "Flies on the Water," *Vermont Outdoors Magazine* (June 1999): 23; Hentcy, "The Sporting Naturalist," *Vermont Outdoors Magazine* (July 1999): 17; Kilborn, "Swarms of Dying Mayflies," *The New York Times,* web site; Usinger, *The Life of Rivers and Streams,* various.
17. Kilborn, "Swarms of Dying Mayflies," *The New York Times,* web site.
18. Usinger, *The Life of Rivers and Streams,* 96.
19. Kirn, *Evaluation of Wild Trout Populations in Vermont Streams* (2000), 1.
20. Ellwood Colahan, *The Angler Club Bulletin,* as quoted by Migdalski, *Angler's Guide,* 189.
21. Vermont Department of Fish and Wildlife, *Vermont Management Plan for Brook, Brown and Rainbow Trout,* 10.
22. Migdalski, 157–58.
23. Langdon et al., "Land Ice and Fish," *passim.*
24. Covington, "The State of Fish and Wildlife," 23.
25. Chapman, "Brook Trout," 16–17.
26. Migdalski, *Angler's Guide,* 182–85.
27. Vermont Department of Fish and Wildlife, *Vermont Management Plan,* 11.
28. Ibid., 32.
29. Ibid., 32–33; Appendix I.A, Fig. 2.

30. Camman, *Fishing Vermont's Lakes and Streams,* 93.
31. Estrin, *Montpelier Rivers Report,* 35–36.
32. Allen et al., *Vermont Trout Streams,* 106.
33. Langdon et al., "Land Ice and Fish," fish sampling data.
34. Ring, "Biologists Find Eggs of Sturgeon in Vermont," 3.
35. Kirn, *A Ten-Year Evaluation of the Effects of Ski Area Development on Wild Brook Trout Populations in Vermont Streams,* 1–2.

15. The Long March and a Midnight Alarm (pp. 113–17)

1. Vermont Agency of Environmental Conservation, *Vermont Rivers Study,* 17.
2. Hill, *The Winooski,* 118.
3. Vermont Agency of Environmental Conservation, *Vermont Rivers Study,* 124.
4. Hill, *The Winooski,* 121–23.
5. Pettengill, *The Yankee Pioneers,* 37–67.
6. Nathan Merrill as quoted by Ehlers, "Vermont Outdoors Destination," 13.
7. Pettengill, *The Yankee Pioneers,* 64.
8. Ibid., 66–67.

16. The Farmers (pp. 118–29)

1. Czaplinski, "Learning to Keep the Mad River Happy," 7.
2. Fallon and Enig, "Splendor from the Grass."
3. Murphy, *Greener Pastures on Your Side of the Fence.*
4. Ibid., 15.
5. Ibid., 14.
6. Ibid., 19ff.
7. Ibid., 86–87.
8. Ibid., 88.
9. Ibid., 142.
10. Fallon and Enig, "Splendor from the Grass," 5.
11. The literature abounds with examples of increased profitability. For a restrained view, the result of a Pennsylvania survey, see Natural Resources Conservation Service, "Farmer Profitability with Intensive Rotational Stocking."
12. Ovid, *The Metamorphoses,* 33–34.
13. Hanson, *Fields without Dreams,* 123–24.
14. Savory, "Four Fundamental Processes that Drive Our Ecosystem."
15. Savory, "Origins of Holistic Management."
16. Nation, *The Stockman Grass Farmer,* 1.
17. Northeast Organic Farming Association of Vermont, "Directory 2002," 2.
18. Murphy, *Greener Pastures on Your Side of the Fence,* 35–36.
19. SARE, "Profitable Dairy Options."
20. Center for Sustainable Agriculture, "Grass Farming."
21. Lowdermilk, "Conquest of the Land."

22. SARE, "Profitable Dairy Options," 3.
23. Nation, *The Stockman Grass Farmer,* 1.
24. Harris, "Connecting Land and Communities," 4.
25. Fallon and Enig, "Splendor from the Grass," 7.
26. *Confession and Covenant of the Congregational Church of Cabot, Vt.*

17. From Gorge to Gorge (pp. 130–35)

1. Vermont Water Quality Division, *The Draft Preliminary Comprehensive River Plan,* 88. Hartland's Quechee Gorge is identified as the largest.
2. Winooski Valley Park District, *Canoe and Natural History Guide,* 34.
3. Weidensaul, *The Raptor Almanac,* 78. The other turkey vulture information is also from this book.
4. Ibid., 169.
5. Ibid., 4–8.
6. Ibid., 8.
7. Ibid., 90.
8. Carlisle, *Look Around Essex and Williston, Vermont,* 10.

18. The Bobcat Man (pp. 136–48)

1. Rolley, "Bobcat," 678.
2. Pete Quinlan, "Outdoor Pointers," 45.
3. Rolley, "Bobcat," 671.
4. Ibid., 675.
5. There is a controversy among scientists over "competition" in the natural world, a minority emphasizing "cooperation" as the more fundamental character of relationships among species. Since this is an argument into which an amateur steps at his peril, I will simply refer the reader to Robert Augros and George Stanciu's *The New Biology: Discovering the Wisdom in Nature.*
6. Rolley, "Bobcat," 675.
7. Levin, "Of Catamounts and Panthers," 23.

19. The Haunted House (pp. 149–57)

1. Vermont Agency of Environmental Conservation, *Vermont Rivers Study,* 47. Note that this was calculated as of 1986. "In Vermont an urbanized area is defined as an incorporated community and adjacent settled areas that together have a minimum population of 5,000 people" (45).
2. Carlisle, *Look Around Essex and Williston, Vermont,* 14–15.
3. Ibid., 49.
4. Whitcomb information comes from conversation and from Chiera, "The Whitcomb Farm," and *The Essex Reporter.*

5. Carlisle, *Look Around Essex and Williston, Vermont,* 48.
6. Ibid., 49.
7. Barg, *The Streamside Sentinel,* 8.
8. For an early account of the project by the soil conservationist in charge of vegetation work for Project Vermont, see Altpeter, "Use of Vegetation," 99–107.
9. Short and Ryan, *The Winooski River Watershed Evaluation Project Report,* 2–3, 16, 18, 67.
10. Ibid., 7–14.
11. Ibid., 62–63, 77.
12. Ibid., 74.
13. Ibid., 61.
14. NRCS Watershed Science Institute, *Examining a 1930's Restoration of the Winooski River Watershed, Vermont,* 1, 3, 8.

20. Power (pp. 158–68)

1. Doland, *Hydro Power Engineering,* 3–11.
2. Merriman, *Treatise on Hydraulics,* 396.
3. National Hydropower Association, "Facts You Should Know about Hydropower," "A Clean Energy Source for Our Future," "9 Myths about Hydropower."
4. Anisfield, "Habitat and Hydropower," 26–27.
5. Vermont Agency of Environmental Conservation, *Vermont Rivers Study,* 19.
6. National Hydropower Association, "9 Myths about Hydropower."
7. International Rivers Network, web site; China Highlights, web site.
8. Merriman, *Treatise on Hydraulics,* 434.
9. Waterwheel Factory, "The Poncelet Design."
10. Waterwheel Factory, "The Fourneyron Design."

21. The Abenaki (pp. 169–79)

1. Calloway, *The Western Abenakis of Vermont,* xvi; *Brattleboro Reformer* (Sept. 26, 2002), p. 3.
2. I learned much from the following: Haviland and Power, *The Original Vermonters;* Wiseman, *The Voice of the Dawn;* Calloway, *The Western Abenakis of Vermont;* Foster and Cowan, *In Search of New England's Native Past;* Gallagher, *Breeding Better Vermonters;* and Jacobs-Carnahan, *The State of Vermont's Response.* For broader accounts of Indian experience in the Northeast, see Jennings, *The Invasion of America;* Axtell, *The European and the Indian;* and Cronon, *Changes in the Land.*
3. Haviland and Power, *The Original Vermonters,* 155–56.
4. Ibid., 260, 159, 247. The 10,000 figure is from Moody, "Abenaki Nation."
5. Haviland and Power, *The Original Vermonters,* 186–97; Day, "The Eastern Boundary of Iroquoia"; Calvin Martin, *Keepers of the Game;* Krech, *The Ecological Indian* and *Indians, Animals, and the Fur Trade;* Cronon, *Uncommon*

Ground; McKibben, *The End of Nature;* Hamell, "Strawberries"; Speck, "Penobscot Tales"; Whitehead, *Stories from the Six Worlds;* and many titles by Joseph Bruchac, a contemporary Abenaki storyteller.

6. Haviland and Power, *The Original Vermonters,* 204.
7. Ibid., 158–68.
8. Day, "The Eastern Boundary," 25.
9. For a similar argument, see Albers, *Hands on the Land,* 63.
10. Hamell, "Strawberries," 77.
11. Calvin Martin, *Keepers of the Game,* chapter 5, but found widely.
12. Speck, as cited in ibid., 113–14.
13. Ibid., 15.
14. Haviland and Power, *The Original Vermonters,* 186–97; Day, "The Eastern Boundary," 120–22.
15. Day, "The Eastern Boundary," 120–21.
16. Haviland and Power, *The Original Vermonters,* 177.
17. Krech, *The Ecological Indian,* 122. In *Keepers of the Game,* Martin argued that Indians in Eastern Canada thought that European diseases were caused by animals that were thereby violating the previously existing respectful relationship between animal and human. Moreover, the shamans couldn't cure the diseases. Through the fur trade, the Indian not only obtained European goods but also could take revenge on his animal enemies. This argument was criticized extensively in Krech, *Indians, Animals, and the Fur Trade.*

22. *A City of Mills: The Superintendent and the Doctor (pp. 180–93)*

1. Jellison, *Ethan Allen,* vii.
2. Ibid., 62.
3. Nathan Perkins, 1789, as found in Bellesiles, *Revolutionary Outlaws,* 257.
4. Jellison, *Ethan Allen,* 18ff.
5. Ibid., 3.
6. Ibid., 75–83.
7. Ibid., 78.
8. Ibid., 79–80. Over the years, the community was known as the Falls, Colchester Falls, Burlington Falls, Winooski Falls, and Winooski. Incorporated as a village within the town of Colchester in 1866, it became an independent city—a very small one of just 1.43 square miles—in 1922 (Feeney, *The Great Falls on Onion River,* 7, 122).
9. For possible early French place names, building remains, and land use patterns, see Meeks, *Time and Change in Vermont,* 10ff.
10. Bellesiles, *Revolutionary Outlaws,* 283.
11. Feeney, *The Great Falls on Onion River,* chapter 2.
12. *Northern Sentinel,* February 11, 1814, as found in Graffagnino, *Ethan and Ira Allen,* v. 1, p. xxii, n. 23.
13. Blow, "Industrial Growth in Winooski," 30ff.
14. Hope, "The Fuller's Teasel," 42ff.
15. Krawitt, *Clickety Clack,* 31.

16. Blow, "Industrial Growth in Winooski," 34.
17. Feeney, *The Great Falls on Onion River,* 37.
18. Blow, "Industrial Growth in Winooski," 39.
19. Feeney, "The Great Falls on Onion River," 59.
20. This and the following paragraph are mainly from Versteeg. Versteeg also refers the reader to Suzanne Duncan, "The Role of the Merino in the Sheep and Woolen Industries of Vermont," Special Collections, University of Vermont.
21. Slaybaugh, "Why the Mills Closed."
22. For child labor, see Gillies, "Regulating Child Labor in Vermont." Rita Martel also tells the story in the video, *The Woolen Mills,* Winooski Historical Society.
23. Hareven and Langenbach, *Amoskeag.*
24. See Krawitt, *The Mills at Winooski Falls,* Section Four, "Many Languages, Many Cultures."
25. Roy, "Mill Superintendent and Safety Director," and, mainly, my conversation with him.
26. Feeney, *The Great Falls on Onion River,* 109.
27. Moran, *The Belles of New England,* 115.
28. O'Brien, "An Interview," and, mainly, my conversation with him.
29. Feeney, *The Great Falls on Onion River,* 141–44.
30. "Winooski by the Numbers," *Seven Days;* Larson, "State Sees Decline."

23. *To the Lake (pp. 194–99)*

1. For information about the Intervale, I have drawn on Natural Assets Consultants, *The Burlington Intervale;* Lawrence, *Intervale Nature Guide;* and Reno, "The Burlington Intervale."
2. Gallagher, *Breeding Better Vermonters.*
3. Dann, "Where the Wild Things Are," 194–98.
4. Daniel and Ward, *Where the Wild Things Are.*
5. Grogan, "At the Intervale." For more Intervale information, see Makinson, "Reestablishing Farming"; Intervale Foundation, "Intervale Farm Program"; and Rathke, "Intervale."
6. Hill, *The Winooski,* 127–32.

24. *The Trapper (pp. 200–212)*

1. Quotations in this chapter to this point are from Maurer, "Waterfront Living."
2. "Outdoor News," *Vermont Outdoors Magazine:* 24.
3. Manitoba, "Hudson's Bay Company Archives."
4. McGee, "The Use of Furbearers," 13.
5. Outwater, *Water: A Natural History,* 8. See also McGee, "The Use of Furbearers"; and Ray, "The Fur Trade."
6. Outwater, *Water: A Natural History,* 6.

7. Ray, "The Fur Trade," 26.
8. Giroux, "The Role of the Trapper Today," 56.
9. Bice, "A Trapper Remembers," 52, 54.
10. Vermont Department of Fish and Wildlife, "Featured Species: Marten (*Martes americana*)," *Vermont Furbearer Management Newsletter* (Fall/Winter 2001):6.
11. Northeast Furbearer Resources Technical Committee, *Trapping and Furbearer Management*, 33.
12. This general information may be found in many places. Two convenient sources are the Northeast Furbearer Resources Technical Committee, and Vermont Department of Fish and Wildlife, *The Role of Regulated Trapping*.

25. Learning to See (pp. 213–27)

1. Shedd, *Owls Aren't Wise*, 59–60.
2. Rezendes, *The Wild Within*, 99.
3. Ibid., 82–85.
4. Ibid., 99–103.
5. Some naturalists add that the new logging practice of clear-cutting in these forests after World War II, made possible by modern machinery, triggered the rapid return of the moose, for clear-cut areas growing back produced "the young hardwood trees which are prime moose fodder." Shedd, *Owls Aren't Wise*, 287.
6. Except for Project Elm, the following details of her early life are from Wolkomir, "Following the Footsteps of Fox and Bear." Project Elm information is from Dickerson and Chase, "Susan Morse and her 'Project Elm'."
7. Rezendes, "Paul Rezendes: Photographs and Nature Programs."
8. Rezendes, *The Wild Within*, 5.
9. Ibid., chapter 1, "The Web of Life."
10. Ibid., 211–12.
11. Ibid., 43–45.
12. Ibid., 71.

Published Sources

Albers, Jan. *Hands on the Land: A History of the Vermont Landscape.* Cambridge: MIT Press, 2000. Published for the Orton Family Foundation, Rutland, Vermont.

Allen, Farrow, and Edward Antczak, and Peter Shea. *Vermont Trout Streams.* Burlington, Vt.: Northern Cartographic, Inc., 1985.

Altpeter, L. Stanford. "Use of Vegetation in Control of Streambank Erosion in Northern New England," *Journal of Forestry* (February 1944): 99–107.

Anisfield, Nancy. "Habitat and Hydropower," *Vermont Outdoors Magazine* (June 1999): 26–27.

Augros, Robert, and George Stanciu. *The New Biology: Discovering the Wisdom in Nature.* Boston: New Science Library, 1987.

Axtell, James. *The European and the Indian: Essays in the Ethnohistory of Colonial North America.* Oxford: Oxford University Press, 1981.

Barg, Lori, editor and researcher for the Vermont Agency of Natural Resources. *The Streamside Sentinel,* n.d.

Barg, Lori, and George Springston. "Floods and the Great Brook in Plainfield, Vermont," *The Upper Winooski Field Naturalists Contributions Series,* ed. Arthur V. Gilman and Sarah Gilbert (No. 2, June 2001).

Basilchuk, Nancy. "Most Lake Pollution Comes from Yards," *The Burlington Free Press,* March 24, 2002, 1A.

Bellesiles, Michael A. *Revolutionary Outlaws: Ethan Allen and the Struggle for Independence on the Early American Frontier.* Charlottesville: University Press of Virginia,1993.

Bernier, Dicky. "Maine Whitetail Connection," *Outdoors Magazine* (March 2001): 19.

Berry, Wendell. *Life is a Miracle: An Essay Against Modern Superstition.* Washington, D.C.: Counterpoint, 2001.

Bice, Ralph. "A Trapper Remembers," in *Wild Furbearer Management and Conservation in North America,* ed. Milan Novak et al., 51–54.

Blais, Tom. "The Buck Stops Here," *Outdoors Magazine* (July 2002): 28.

Blaisdell, Harold. *The Philosophical Fisherman.* New York: Nick Lyons Books, 1986. Originally published Boston: Houghton Mifflin, 1969.

Blow, David J. "Industrial Growth in Winooski," in *The Mills at Winooski Falls,* ed. Laura Krawitt, 30–41.

Bruchac, Joseph. *Dawn Land.* Golden, Colo.: Fulcrum Publishing, 1993.

———. *The Waters Between: A Novel of the Dawn Land.* Hanover, N.H.: University Press of New England, 1998.

Buck, John. "The State of Fish and Wildlife: View from the Vermont Fish and Wildlife Department," *Vermont Outdoors Magazine* (August 1998): 20; (December 1999): 6.

Cabot Creamery. <www.cabotcheese.com>.

Cairns, Bill. "Flies on the Water," *Outdoors Magazine* (October 2002): 17; (July 2002): 15; *Vermont Outdoors Magazine* (August 1999): 23; (June 1999): 23.

Calloway, Colin G. *The Western Abenakis of Vermont, 1600–1800: War, Migration, and the Survival of an Indian People.* Norman: University of Oklahoma Press, 1994. Published in 1990 in the Civilization of the American Indian Series.

Camman, Peter F. *Fishing Vermont's Lakes and Streams: A Guide to the Green Mountain State's Best Trout and Bass Waters.* Woodstock, Vt.: Backcountry Publications, 1992.

Carlisle, Lilian Baker, ed. *Look Around Essex and Williston, Vermont.* Heritage Series Pamphlet Number Four. Burlington, Vt.: Chittenden County Historical Society, 1973.

Center for Sustainable Agriculture, University of Vermont. "Grass Farming: Grazing the Vermont Landscape for Profitable Farming," <www.uvm.edu/~susagctr/grass.html>.

———. "Second Annual Grazing Conference Draws a Crowd," *Cultivating Connections* (Spring 1998): 5 (newsletter).

Chapman, Liz. "Brook Trout: Everybody's Favorite Vermonter," *Vermont Outdoors Magazine* (July 1998): 16–17.

Chiera, Ann Della. "The Whitcomb Farm," *Vermont Life* (Spring 1956): 52–57.

China Highlights. <www.chinahighlights.com/yangtzecruise/faq/facts.htm>.

Columbia Encyclopedia. 5th ed. New York: Columbia University Press, 1993.

Confession and Covenant of the Congregational Church of Cabot, Vt. Montpelier, Vt.: E. P. Walton, Printer, 1859.

Connell, Kinny. "Transition: Message from the President," *Mad River Matters* (Fall–Winter 2000).

Cousins, Freddie. "Opportunity for Change: Relicensing Waterbury Hydroelectric," *Currents* (Newsletter of the Friends of the Winooski River) (Spring 2000): 1, 7.

Covington, Scott. "The State of Fish and Wildlife: View from the Vermont Fish and Wildlife Department," *Outdoors Magazine* (November 2001): 23.

Cronon, William. *Changes in the Land: Indians, Colonists, and the Ecology of New England.* New York: Hill & Wang, 1983.

———, ed. *Uncommon Ground: Rethinking the Human Place in Nature.* New York: W. W. Norton & Company, 1996.

Czaplinski, Richard. "Learning to Keep the Mad River Happy," *Mad River Matters* (Fall/Winter 2000): 1, 7.

Daniel, Alicia, and Mark Ward. *Where the Wild Things Are: Large Mammal Habitat and Corridors in Burlington, Vermont,* prepared for the Winooski Valley Park District, October 2002.

Dann, Kevin. "Where the Wild Things Are," in *Intervale Nature Guide,* by Gale Lawrence, 194–98.

Dathe, David. *Fundamentals of Historical Geology.* Dubuque, Iowa: Wm. C. Brown Publishers, 1993.

Day, Gordon M. "The Eastern Boundary of Iroquoia: Abenaki Evidence," in *In Search of New England's Native Past,* ed. Michael K. Foster and William Cowan, 116–22.

———. "The Western Abenaki Transformer," in Foster and Cowan, 183–94.

———. "Western Abenaki," in Foster and Cowan, 202–22.

Deer Management Team, Vermont Department of Fish and Wildlife. "Deer Wintering Areas and Development in Vermont," February 1993.

———. White-Tailed Deer Management Plan for the State of Vermont (1997–2006,) July 1997.

Dick, H. Lenox H. *The Art and Science of Fly Fishing.* Princeton, N.J.: D. Van Nostrand Company, Inc., 1966.

Dickerson, Mary Jane, and Joan Strausbaugh Chase. "Susan Morse and her 'Project Elm,'" *Vermont Life* (Spring 1976): 59–61.

Doland, James J. *Hydro Power Engineering: A Textbook for Civil Engineers.* New York: The Ronald Press Company, 1954.

Dwight, Timothy. *Travels in New England and New York,* 4 vols. Barbara Miller Soloman, ed., with assistance of Patricia M. King. Cambridge, Mass.: Harvard University Press, 1969. First published posthumously, New Haven, Conn., 1821.

Ehlers, James. "Vermont Outdoors Destination: Winooski River Middlesex to Bolton," *Vermont Outdoors Magazine* (July 1998).

Eiseley, Loren. *The Firmament of Time.* New York: Atheneum, 1966.

Estrin, Hans et al. *Montpelier Rivers Report,* prepared for the Montpelier Conservation Commission as part of the Montpelier River Project, March 1995.

Fallon, Sally, and Mary G. Enig. "Splendor from the Grass," the Weston A. Price Foundation, <www.westonaprice.org/farming/splendor.html>.

Faulkner, Robert K. *Francis Bacon and the Project of Progress.* Lanham, Md.: Rowman & Littlefield Publishers, Inc., 1993.

Feeney, Vincent Edward. *The Great Falls on Onion River: A History of Winooski, Vermont.* Winooski, Vt.: The Winooski Historical Society, 2002.

Flannery, Tim. *The Eternal Frontier: An Ecological History of North America and Its Peoples.* New York: Atlantic Monthly Press, 2001.

Foster, Michael K., and William Cowan, eds. *In Search of New England's Native Past: Selected Essays by Gordon M. Day.* Amherst: University of Massachusetts Press, 1998.

Friends of the Mad River. "Friends of the Mad River," an undated brochure; "2002 Mad River Watch Sampling Results," <www.friendsofthemadriver.org>.

Friends of the Mad River and Mad River Valley Planning District. *The Best River Ever: A Conservation Plan to Protect and Restore Vermont's Beautiful Mad River Watershed.* Waitsfield, Vt., 1995.

Gallagher, Nancy L. *Breeding Better Vermonters: The Eugenics Project in the Green Mountain State.* Hanover, N.H.: University Press of New England, 1999.

Gillies, Paul. "Regulating Child Labor in Vermont," in *The Mills at Winooski Falls,* ed. Laura Krawitt, 153–70.

Gilluly, James, Aaron C. Waters, and A. O. Woodford. *Principles of Geology.* San Francisco: W. H. Freeman & Company, 1954.

Giroux, Alcide. "The Role of the Trapper Today," in *Wild Furbearer Management and Conservation in North America,* ed. Milan Novak et al., chapter 6.

Gow, Mary. "Stabilizing the Streambank," *Mad River Matters* (Winter 1996–1997), a publication of the Friends of the Mad River.

Graffagnino, J. Kevin, ed. *Ethan and Ira Allen: Collected Works,* 3 vols. Benson, Vt.: Chaldize Publications, 1992.

———. "'The Country My Soul Delighted In': The Onion River Land Company and the Vermont Frontier," *New England Quarterly* (March 1992): 24–60.

Grimshaw, Paul, & Son. *Grimshaws Trapper Supply Catalog* (2001–2002 Seasons), Champlain, N.Y.

Grogan, John. "At the Intervale in Vermont, a Community Turns a Forgotten Wasteland into a Thriving Organic Mecca," from *Organic Gardening Magazine,* published on the Intervale Foundation web site, <www.intervale.org/Paradise.html>.

Guertin, David. "Flies for Vermont Trout Streams," <http://caddis.middlebury.edu/trout>.

Halls, Lowell K. *White-Tailed Deer: Ecology and Management.* A Wildlife Management Institute Book. Harrisburg, Penn.: Stackpole Books, 1984.

Hamell, George R. "Strawberries, Floating Islands, and Rabbit Captains: Mythical Realities and European Contact in the Northeast During the Sixteenth and Seventeenth Centuries," *Journal of Canadian Studies* (April 1987): 72–94.

Hanson, Victor Davis. *Fields without Dreams: Defending the Agrarian Idea.* New York: The Free Press, 1996.

Hareven, Tamara, and Randolph Langenbach. *Amoskeag: Life and Work in an American Factory City.* Hanover, N.H.: University Press of New England, 1978.

Harris, Gwyneth. "Connecting Land and Communities," *Cultivating Connections* (Autumn 2001): 4. Newsletter of the Center for Sustainable Agriculture, the University of Vermont.

Hausman, Bob. "Rock Collector Takes His Hobby Seriously," *The Caledonian Record* (1997).

Haviland, William A., and Marjory A. Power. *The Original Vermonters: Native Inhabitants, Past and Present.* Hanover, N.H.: University of Vermont—University Press of New England, rev. ed., 1994.

Hentcy, Kathleen. "The Sporting Naturalist," *Vermont Outdoors Magazine* (December 1998): 28; (July 1999): 17.

Hickoff, Steve, and Rhey Plumley. *Steve Hickoff and Rhey Plumley's Flyfisher's Guide to Northern New England: Vermont, New Hampshire, and Maine.* Belgrade, Mont.: Wilderness Adventures Press, 1999.

Hill, Ralph Nading. *The Winooski: Heartway of Vermont.* Rivers of America. New York: Rinehart & Company, Inc., 1949.

Holland, Dan. *The Trout Fisherman's Bible.* Garden City, N.Y.: Doubleday & Company, Inc., 1949.

Hope, Peter. "The Fuller's Teasel," in *The Mills at Winooski Falls,* ed. Laura Krawitt, 42–44.

Howe, Jeff. "The Charlotte Whale: A 12,500–10,000-Year-Old Post-Glacial Ancestor of White (Beluga) Whales Found in Charlotte, Vermont in 1849," February 1993, an account distributed at the University of Vermont's Perkins Museum of Geology.

Hutton, James. "Theory of the Earth; or an Investigation of the Laws Observable in the Composition, Dissolution and Restoration of the Land Upon the Globe," *Transactions of the Royal Society of Edinburgh* (1788).

Imbrie, John, and Katherine Palmer Imbrie. *Ice Ages: Solving the Mystery.* Cambridge, Mass.: Harvard University Press, 1979.

International Rivers Network. <http://iso.hrchina.org:8151/old_site/reports/3gorges.html>; <www.irn.org/programs/threeg/>.

Intervale Foundation. "Intervale Farm Program Incubator Program," <www.intervale.org/IncubatorProgram.html>.

Jacobs, Elbridge Churchill. *The Physical Features of Vermont,* Vermont Geological Survey, Montpelier, Vt.: Vermont State Development Department, February 1950.

Jacobs-Carnahan, Eve (Special Assistant Attorney General; William H. Sorrell, Attorney General). *The State of Vermont's Response to Petition for Federal Recognition of the St. Francis-Sokoki Band of the Abenaki Nation of Vermont,* December 2002.

Jennings, Francis. *The Invasion of America: Indians, Colonialism, and the Cant of Conquest.* New York: W. W. Norton, 1976. Published in 1975 by University of North Carolina Press.

Jellison, Charles A. *Ethan Allen: Frontier Rebel.* Syracuse: Syracuse University Press, 1969.

Johnson, Charles W. *The Nature of Vermont: Introduction and Guide to a New England Environment.* Hanover, N.H.: University Press of New England, 1980.

Keating, Ted, ed. *Trout Fishing.* The Sports Afield Library. New York: Thomas Nelson & Sons, 1962.

Kilborn, Peter T. "Swarms of Dying Mayflies Bring Good News," *The New York Times,* <www.nytimes.com/2002/08/05>.

Kirn, Rich. *Evaluation of Wild Brook Trout Populations in Vermont Streams* Vermont Department of Fish & Wildlife, 2000.

———. *A Ten-Year Evaluation of the Effects of Ski Area Development on Wild Brook Trout Populations in Vermont Streams.* Vermont Department of Fish and Wildlife, July 1997.

Klyza, Christopher McGrory, and Stephen C. Trombulak. *The Story of Vermont: A Natural and Cultural History.* Hanover, N.H.: Middlebury College Press—University Press of New England, 1999.

Knighton, David. *Fluvial Forms and Processes: A New Perspective.* London: Arnold and Oxford University Press, 1998.

Krawitt, Laura, ed. *The Mills at Winooski Falls: Winooski and Burlington, Vermont.* Winooski, Vt.: Onion River Press, 2000.

———, ed. *Clickety Clack: Wool and Waterpower: An Interdisciplinary Curriculum Guide for the Heritage Winooski Mill Museum.* Burlington, Vt., 2000. Compiled by Nancy Keller, Claire Myers, and Martha Olbrych.

Krech, Shepard, III, ed. *Indians, Animals, and the Fur Trade: A Critique of Keepers of the Game.* Athens, Ga.: University of Georgia Press, 1981.

———. *The Ecological Indian: Myth and History.* New York: W. W. Norton & Company, 1999.

Langdon, Rich, Mark Ferguson, and Ken Cox. "Land Ice and Fish," a manuscript, courtesy of the authors, later to be published.

Lange, Willem. *Wildlife Journal.* New Hampshire Public Television, October 18, 2002.

Larsen, Frederick D. "Glacial History of the Montpelier, Vermont, 7.5 Minute Quadrangle," in *New England Intercollegiate Conference Guidebook,* No. 91, ed. S. F. Wright, 286–300.

———. "History of Glacial Lakes in the Dog River Valley, Central Vermont," Field Trip B-4, *Guidebook for Field Trips in Vermont,* vol. 2, New England Intercollegiate Geological Conference, 1987.

Larson, Krista. "State Sees Decline in Number of French-speaking Residents, *Brattleboro Reformer,* August 6, 2002, 1, 8.

Lawrence, Gale. *Intervale Nature Guide,* prepared for the Intervale Foundation, n.d.

Levin, Ted. "Of Catamounts and Panthers," *Vermont Life* (Spring 1997): 22–25.

Lowdermilk, W. C. "Conquest of the Land through Seven Thousand Years," Agriculture Information Bulletin No. 99, U.S. Department of Agriculture Soil Conservation Service, 1953, revised 1975.

Lutgens, Frederick K., and Edward J. Tarbuck. *Essentials of Geology,* 4th ed. New York: Macmillan Publishing Company, 1992.

Lyell, Charles. *Principles of Geology,* James A. Secord, ed. London: Penguin, 1997. An abridgment of *Principles of Geology Being an Attempt to Explain the Former Changes of the Earth's Surface By Reference to Causes Now in Operation,* first published in London, 3 vols., 1830–1833.

Makinson, Daphne. "Reestablishing Farming," the Intervale Foundation, <www.intervale.org/reestablishing.html>.

Manitoba, Province of. "Hudson's Bay Company Archives," <www.gov.mb.ca/chc/archives/hbca/about/hbca.html>.

Marsh, George Perkins. *Man and Nature; or, Physical Geography as Modified by Human Action* (1864), David Lowenthal, ed. Cambridge, Mass.: Harvard University Press, 1965.

Martin, Calvin. *Keepers of the Game: Indian-Animal Relationships and the Fur Trade.* Berkeley: University of California Press, 1978.

Martin, J. M., Project Leader. *Vermont Stream Survey 1952–1960.* Montpelier, Vt.: State of Vermont Fish and Game Department, 1962.

Maurer, Carol. "Waterfront Living," *The Sunday Rutland Herald and Times Argus* December 14, 1997.

McCabe, Richard E., and Thomas R. McCabe. "Of Slings and Arrows: An Historical Perspective," in *White-tailed Deer,* ed. Halls, chapter 2.

McGee, Harold F., Jr. "The Use of Furbearers by Native North Americans After 1500," in *Wild Furbearer Management and Conservation in North America,* ed. Milan Novak et al., chapter 2.

McKibben, Bill. *The End of Nature*. New York: Random House, 1989.

Meeks, Harold A. *Time and Change in Vermont: A Human Geography.* Chester, Conn.: Globe Pequot Press, 1986.

———. *Vermont's Land and Resources*. Shelburne, Vt.: New England Press, 1986.

Merriman, Mansfield. *Treatise on Hydraulics,* 10th ed, rev. with the assistance of Thaddeus Merriman. New York: John Wiley & Sons, Inc., 1930.

Migdalski, Edward C. *Angler's Guide to the Fresh Water Sport Fishes*. New York: Ronald Press Company, 1962.

Moody, John. "Abenaki Nation," <www.abenakination.org/history.html>.

Montpelier (Vermont), City of. "Wastewater Treatment Facilities," n.d. A municipal brochure.

Moran, William. *The Belles of New England: The Women of the Textile Mills and the Families Whose Wealth They Wove*. New York: Thomas Dunne Books, St. Martin's Press, 2002.

Murphy, Bill. *Greener Pastures on Your Side of the Fence: Better Farming with Voisin Management Intensive Grazing*. 3rd ed. Colchester, Vt.: Arriba Publishing, 1994.

Nation, Allan. *The Stockman Grass Farmer.* Special Catalog Edition, n.d.

National Environmental Education and Training Foundation and Roper Starch Worldwide. "Myths and Facts about Water," *The National Report Card on Environmental Knowledge, Attitudes and Behaviors,* <www.rivernetwork .org>.

National Hydropower Association. "Facts You Should Know About Hydropower," <www.hydro.org/hydrofacts.asp>; "A Clean Energy Source for Our Future," <www.hydro.org/hydrofacts/future.asp>; "9 Myths About Hydropower," <www.hydro.org/hydrofacts/myths.asp>.

Natural Assets Consultants (Mike Stevens, Judith Rubin, Danielle Livellara, and Jason Rochette). *The Burlington Intervale: Connecting Natural Areas and Agriculture,* prepared for the Intervale Foundation, May 1995.

Natural Resources Conservation Service, United States Department of Agriculture. "Farmer Profitability with Intensive Rotational Stocking: Better Grazing Management for Pastures," n.d.

Neill, Maudean. *Fiery Crosses in the Green Mountains: The Story of the Ku Klux Klan in Vermont*. Randolph Center, Vt.: Greenhills Books, 1989.

Nelson, Richard. *Heart and Blood: Living with Deer in America*. New York:Vintage Books, 1998.

Northeast Furbearer Resources Technical Committee. *Trapping and Furbearer Management in North American Wildlife Conservation,* 2001.

Northeast Organic Farming Association of Vermont. "Directory 2002."

Novak, Milan, James A. Baker, Martyn E. Obbard, and Bruce Mallock, eds. *Wild Furbearer Management and Conservation in North America*. The Ontario Trappers Association and the Ontario Ministry of Natural Resources, 1987.

NRCS Watershed Science Institute. *Examining a 1930's Restoration of the Winooski River Watershed, Vermont,* a Case Summary Report of the Institute, University of Vermont, n.d.
O'Brien, Robert E., M.D. "An Interview: John O'Brien, Former General Manager, Recollections of Robert E. O'Brien, M.D.," in *The Mills at Winooski Falls,* ed. Laura Krawitt, 148–52.
"Outdoor News." *Vermont Outdoors Magazine,* January 1999: 24.
Outwater, Alice. *Water: A Natural History.* New York: Basic Books, 1996.
Ovid. *The Metamorphoses,* tr. Horace Gregory. New York: New American Library, 1960.
Pease, Jim. "Setting the Standards: Watershed Planning and TMDL's," *Currents: Newsletter of the Friends of the Winooski River* (Fall 2000).
Pettengill, Samuel B. *The Yankee Pioneers: A Saga of Courage.* Rutland, Vt.: Charles E. Tuttle Company, 1971.
Pitkin, Caleb, ed. *Cabot, Vermont: A Collection of Memories from the Century Past.* Cabot, Vt.: Cabot Oral History Committee, 1999.
———, ed. *Marshfield, Vermont: A Photographic Album 1860–1930.* Marshfield, Vt.: Marshfield Historical Society, 1998.
Power, Marjory W., and James B. Petersen. *Seasons of Prehistory: 4000 Years at the Winooski Site.* Montpelier, Vt.: State Division of Historic Preservation, 1984.
Pyne, Lawrence. "Lake Champlain Valley Report," *Vermont Outdoors Magazine* (June 1998): 28.
Pyne, Lawrence, and Marianne Eaton. *Outdoor Journal,* Vermont Public Television, November 3, 2002.
Quick, Jim. *Trout Fishing and Trout Flies.* South Brunswick, N.J.: A. S. Barnes and Co., 1957.
Quinlan, Pete. "Outdoor Pointers," *Outdoors Magazine* (October 2002).
Quinlan, Sharon, Jim Ryan, and Jennifer Shepherd. *Vermont Stream Bank Restoration Guide.* An Americorps project sponsored by the Natural Resources Conservation Services, U.S. Department of Agriculture, November 1996.
Rathke, Lisa. "Intervale: From Farm to Compost to Fuel," *Brattleboro Reformer,* September 27–28, 2003, 26.
Ray, Arthur J. "The Fur Trade in North America: An Overview from a Historical Geographical Perspective," in *Wild Furbearer Management and Conservation in North America,* ed. Milan Novak et al., chapter 3.
Recer, Paul. "Deepest Ice Age Pushed Earth along Evolution." *Brattleboro Reformer,* August 28, 1998.
Reno, Rebecca. "The Burlington Intervale," An Intervale Community Farm Publication, August 1993.
Rezendes, Paul. "Paul Rezendes: Photography and Nature Programs," Spring–Summer 2002, a brochure of offerings.
———. *The Wild Within: Adventures in Nature and Animal Teachings.* New York: Berkley Books, 1999.
———. *Tracking and the Art of Seeing: How to Read Animal Tracks and Sign.* New York: HarperPerennial, 1999.

Ring, Wilson. "Biologists Find Eggs of Sturgeon in Vermont," *Brattleboro Reformer,* May 24–25, 2003, 3.

Rolley, Robert E. "Bobcat," in *Wild Furbearer Management and Conservation in North America,* ed. Milan Novak et al., chapter 50.

Roth, Bernhard A. *The Complete Beginner's Guide to Canoeing.* Garden City, N.Y.: Doubleday & Company, Inc., 1977.

Roy, Raymond. "Mill Superintendent and Safety Director: Recollections of Raymond Roy," in *The Mills at Winooski Falls,* ed. Laura Krawitt, 145–47.

Royar, Kimberly, Christopher Bernier, and Beth Adler, eds. "Bobcat and Fisher DNA Analysis," *Vermont Furbearer Management Newsletter* (Fall/Winter 2002): 9. Vermont Fish and Wildlife Department.

S. Morgan Smith Company. Bulletin (turbines), n.d.

SARE (Sustainable Agriculture Research and Education). "Profitable Dairy Options," n.d. <www.uvm.edu/~nesare/what_is.html>.

Savory, Allen."Four Fundamental Processes that Drive Our Ecosystem," *The Allen Savory Center for Holistic Management,* <www.holisticmanagement.org/ahm_four.efm>.

———. "Origins of Holistic Management: Economic, Social and Environmental Considerations," <www.holisticmanagement.org/ahm_orig.cfm?>

Scanlon, Kyle. "The Biggest and Best," *Vermont Outdoors Magazine* (February 2000).

Schama, Simon. *Landscape and Memory.* New York: Vintage Books, 1996.

Schwiebert, Ernest G., Jr. "The Evolution of Fly Fishing," in *Matching the Hatch: A Practical Guide to Imitation of Insects Found on Eastern and Western Trout Waters.* London: Macmillan Company, 1955, 1–7.

Shedd, Warner. Trudy Nicholson, illustrator. *Owls Aren't Wise and Bats Aren't Blind: A Naturalist Debunks Our Favorite Fallacies about Wildlife.* New York: Harmony Books, 2000.

———. "Think Deer," *New Hampshire Wildlife* (July–August 1985; May–June 1996; July–August 1996).

———. "Wildlife Matters," *Outdoors Magazine* (March 2001): 17; (April 2001): 11; (February 2002): 9.

———. "Wildlife Matters," *Vermont Outdoors Magazine* (March 1998): 6; (August 1999): 7; (July 2000): 12; (November 2000): 14.

Short, Heather, and Jim Ryan (Americorps Members). *The Winooski River Watershed Evaluation Project Report: A Comprehensive Analysis of Project Vermont: A Soil Conservation Service Erosion Control Demonstration Project,* October 1996.

Slaybaugh, Douglas. "Why the Mills Closed," in *The Mills at Winooski Falls,* ed. Laura Krawitt, 178–83.

Smith, Francis. "Test of Survival," *Outdoors Magazine* (February 2002): 30.

Speck, Frank G. "Penobscot Tales and Religious Belief," *Journal of American Folklore* (Janurary–March 1935): 1–107.

Stewart David P., and Paul MacClintock. *The Surficial Geology and Pleistocene History of Vermont.* Vermont Geological Survey, Bulletin No. 31. Montpelier, Vt.: Department of Water Resources, 1969.

Sturtevant, William C. "Animals and Diseases in Indian Belief," in *Indians, Animals, and the Fur Trade,* ed. Shepard Krech III, chapter 8.

Thompson, Zadock. *Natural History of Vermont.* Rutland, Vt.: Charles E. Tuttle Company, 1972. Published by the author in 1853.

Trefil, James. "Evidence for a Flood," *Smithsonian* (April 2000).

U.S. PIRG. "Permit to Pollute: How the Government's Lax Enforcement of the Clean Water Act is Poisoning Our Waters." August 6, 2002. Executive summary of a U.S. PIRG report, <http://uspirg.org>.

Usinger, Robert L. *The Life of Rivers and Streams.* New York: McGraw Hill Book Company, 1967.

Vara, Jon. "In Burlington's Backyard," *Vermont Life* (Autumn 1989): 20–27.

Vermont Agency of Environmental Conservation. *Vermont Rivers Study* (1986).

Vermont Agency of Natural Resources. "Streambank and Lakeshore Vegetation Management Procedure," June 16, 1996.

———. *Stream Geomorphic Assessment Handbook and Database.* Phase 1 written and edited by Mike Kline, Christa Alexander, Staci Pomeroy, George Springston, Barry Cahoon, and Larry Becker; Phase 2 by the same except for Springston. Lori Barg provided an "initial draft outline" for both handbooks.

Vermont Department of Environmental Conservation. *Fluvial Geomorphology: A Foundation for Watershed Protection, Management and Restoration.* January 16, 2001.

Vermont Department of Environmental Conservation, Vermont Department of Agriculture Food and Markets, and U.S. Department of Agriculture, Natural Resources Conservation Service. "Vermont Unified Watershed Assessment, an element of the Clean Water Action Plan." September 1998.

Vermont Department of Fish and Wildlife. "Featured Species: Marten (*Martes americana*)," *Vermont Furbearer Management Newsletter* (Fall/Winter 2001): 6.

———. *A Report of Vermont's 2000 Deer Hunting Seasons.*

———. *A Report of Vermont's 2001 Deer Hunting Seasons.*

———. *The Role of Regulated Trapping and the Management of Furbearers in Vermont Present and Future.* Waterbury, Vt., n.d.

———. *Vermont Management Plan for Brook, Brown and Rainbow Trout.* Waterbury, Vt.: 1993.

Vermont Environmental Protection Agency. <www.epa.gov/305b/98report/vt.pdf>.

Vermont Water Quality Division. *The Draft Preliminary Comprehensive River Plan for the Lower Winooski River.* Waterbury, Vt., July 1991.

———. "Vermont 1998 305(b) Report to the Environmental Protection Agency," <www.epa.gov/305b/98report/vt.pdf>.

———. *Vermont Water Quality Standards.* Effective date, April 21, 1997.

Versteeg, Jennie G. "Counting Sheep and Other Worldly Goods," in *The Mills at Winooski Falls,* ed. Laura Krawitt, 18–29.

Wallach, Steven Jay. "Barbara Carpenter: Fifty Years on the Farm," *Vermont Magazine* 14, no. 1 (January–February 2002): 53–56.

Walton, Izaak. *The Compleat Angler.* New York: Random House (Modern Library, n.d.). From 4th ed., first published in 1653.

Waterwheel Factory. "The Poncelet Design," <www.waterwheelfactory.com/poncelet.htm>.

———. "The Fourneyron Design," <www.waterwheelfactory.com/fourney.htm>.

Weber, Max. "Science as a Vocation" (1919), in *From Max Weber: Essays in Sociology,* ed. and tr. H. H. Gerth and C. Wright Mills. London: Routledge & Kegan Paul Ltd., 1948.

Weidensaul, Scott. *The Raptor Almanac.* New York: Lyons Press, 1996, 2000.

Wessels, Tom. *Reading the Forested Landscape: A Natural History of New England.* Woodstock, Vt.: Countryman Press, 1997.

"Whitcombs in Williston Earn Dairy Farm of the Year." *The Essex Reporter,* October 25, 2001, 12–13.

White, Gilbert. *Natural History of Selborne and Observations on Nature.* 2 vols. New York: D. Appleton & Co., 1895. Originally published in 1789.

Whitehead, Ruth Holmes. *Stories from the Six Worlds: Micmac Legends.* Halifax, Nova Scotia: Nimbus Publishing Limited, 1988.

"Winooski by the Numbers," *Seven Days,* October 22–29, 2003, 29A. A Burlington newspaper.

Winooski Conservation District Local Work Group. *What is Your Connection? Your Winooski Valley Watershed.* February 2002 and <www.vacd.org/winooski>.

Winooski Historical Society. *The Woolen Mills: One Hundred Years of Winooski Reminiscences.* William Hobart and Bernhardt Smyle, co-producers. Hill House & Skeezix Enterprises, 1998.

"Winooski Site Yields 10,000-year-old Tools" (AP), *Brattleboro Reformer,* September 26, 2002.

Winooski Valley Park District. *Canoe and Natural History Guide to the Winooski River Valley,* 2nd ed. Burlington, Vt.

Wisconsin Valley Improvement Company. <www.wvic./com>.

Wiseman, Frederick Matthew. *The Voice of the Dawn: An Auto-History of the Abenaki Nation.* Hanover, N.H.: University Press of New England, 2001.

Wolkomir, Richard. "Following the Footsteps of Fox and Bear," *Smithsonian* 27, no. 10 (January 1997): 34–43.

Wright, J. V. "Archaeological Evidence for the Use of Furbearers in North America," in *Wild Furbearer Management and Conservation in North America,* ed. Milan Novak et al., chapter 1.